PREFACE

This manual is designed to provide an outline of the biology and production practices of small animals. It is not possible to present all the information necessary for a complete understanding of the animal and how its needs might best be met; nor can it be expected that everything written here will be readily understood by all readers. However, sufficient information and instruction for successful animal production is presented at a level hopefully comprehensible to most readers and teachers with the equivalent of a high school education. It will be left to teachers and others to adapt the material to specific, local circumstances, make the information accessible and present it in a manner that would be understood. Similarly, even though some of the information might be more appropriate to large-scale production in the temperate zones, the principles involved can often be adapted to small-scale farming enterprises in tropical areas. This will especially involve improving the level of management of breeding, feeding and control of diseases and the microclimate.

Animals must be selected that are capable of reproducing and producing in the environment that the farmer can provide. Knowing the role played by different nutrients gives a basis for understanding why they must be supplied in the diet. Awareness of the normal, healthy appearance and behaviour of animals is essential to recognising and treating diseases and abnormalities. Knowledge of the structure and function of the various body systems can further serve as a model for better understanding the nutrition, health and well-being of the human organism. Armed with such understanding the farmer is more likely to select and manage his animals for higher, more profitable production. Some suggested learning experiences are outlined to emphasize fundamental aspects of selecting, breeding and managing the different species of animals. This material must be considered in light of local conditions and resources.

The references cited provide valuable data and recommendations for successful breeding and management. Additional information can be obtained from any of the many available textbooks, reference books and the publications of agricultural experiment stations and public and private (business) extension services. No effort has been made to reference all sources of information presented. Much has been drawn from class notes and experiences derived from 33 years of teaching courses in the nutrition, physiology, production and management of animals. The patience and skill of the various experts who reviewed this material is gratefully acknowledged.

The book is divided into two parts. Part One presents scientific principles underlying the successful production of animals. Part Two covers the application of these principles in production practices and learning activities for poultry, rabbits, goats, guinea pigs and pigs.

Any publication meant for an international readership will contain words and phrases not readily recognised or understood by all readers (e.g. table poultry and broiler or roaster). To minimise any such misunderstanding a brief glossary of such terms appears in the appendix.

EGM

Raising Small Animals

Part One
ANIMAL SCIENCE

1

ANIMALS AS CONTRIBUTORS TO HUMAN WELFARE

ANIMALS PRODUCE HUMAN FOOD

Animals supply an important part but are not the major source of food for the families on small farms. Even though animals provide other functions, their primary role is food production. Some consider it wasteful to feed grain and other feedstuffs (some of which might be edible by man) to animals to produce milk, eggs and meat. In developed countries, the extensive feeding of grain to animals is based upon the realities of the market place: farmers grow grain for profit, people prefer diets that include animal products and are willing to pay the extra costs. In these developed countries animals provide two-thirds of the protein, one-third of the calories and half of other nutrients that are consumed.

For those who do not consume animal products, either by choice or necessity, it is possible to design a nutritionally balanced diet from a careful selection of plant products (including some fermented matter). However, strict vegetarians must take care to ensure a balance of essential amino acids (especially methionine and lysine) and vitamins, especially riboflavin (B_2) and cobalamin (B_{12}) and the mineral calcium. Energy and total protein can be readily obtained from cereal grains (wheat, maize), the pulses (beans, peas) and roots (potatoes, cassava). The deficiencies inherent in a vegetable diet can readily be corrected by the inclusion of animal products. Furthermore, availability of animal products allows greater flexibility in selecting a varied diet that is both balanced and palatable.

Aside from the nutritive aspects of food, man gets psychic stimuli from eating. Milk, cheese, eggs, meat and other foods of animal origin not only enhance flavour and palatability, but also have a greater capacity to provide satisfaction and a feeling of well-being and satiety than do foods of only plant origin.

Because of the form and function of man's digestive tract, he cannot efficiently use the complete plant but must select young leaves, tubers and the reproductive portions (seeds and fruit). Except in a limited way, the biggest part of the plant can be eaten by man only after its quality has been improved by first passing it through another animal.

Animals make possible a more complete use of land resources. They can eat the by-products (e.g. bran, brewers' grains) that result when plant matter is converted into a form more desirable for man. Animals add value to crop residues by converting these residues to preferred animal products. Furthermore, the forage crops raised to feed animals are planted in rotation

3

with food crops to increase soil fertility and control plant diseases.

Animals can harvest and use feedstuffs from areas or places where human food crops cannot be grown. They subsist on pasturage and crop aftermath and by grazing roadsides, ditch banks and other places not suitable for tilling. They can also scavenge and convert garbage, weeds and undesirable brush to human food. Herbivores can often subsist, grow and reproduce on non-cultivated plants; the more costly cultivated crops of grain and forage such as maize and lucerne are required only for optimum production. Thus, these animals can augment the home food supply where sufficient land might not be available to produce enough crops.

Herbivores such as goats, sheep and rabbits are relatively free of dependence on a balanced protein feed source (but not of a total protein intake). Consequently, the limiting feed factor in their production is often their energy intake.

The production of milk protein is a more efficient biological process than the production of meat. Also, although higher producing animals eat more, their production is more efficient. However, seeking greater yields from fewer, better animals may run counter to a cultural desire for numbers of live animals with little regard for individual productive performance.

ANIMALS PRODUCE USEFUL FIBRE

Wool, hair, pelts and hides can be used for clothing, protection from the elements and aesthetic adornment.

ANIMALS PRODUCE FUEL AND FERTILISER

Dung can be used for fuel, fertiliser, or as a feed source for worm production. Animals can collect plant materials from wide areas and deposit much of the undigested portion in corrals or pens where it can be gathered.

1. Manure can be used as a fuel where there is a scarcity of firewood and where fossil fuel is expensive. Dung can be either burned directly or put into a fermentation vat to produce methane gas that can be burned for heat.
2. Manure applied as fertiliser improves the tilth and permeability of the soil and supplies nitrogen, phosphate and potash as well. The fertilising value will depend somewhat on what the animals ate.
3. Dung can serve as feed for compost worms such as *Eisenia (Helodrilus) foetida, Lumbricus rubellus* or *L. terrestris*. Such worms can within one year produce up to 200 times their initial weight. Under optimal conditions 100 g of worm mass can be produced in each kg of faeces. An adult worm produces about 1000 young worms annually. A bed of faeces 5 to 10 cm thick can be seeded with 2000 to 4000 adult worms per square metre. After 3 to 4 months the worms can be harvested.

The optimal environmental condition for worms is about 22°C to

26°C and 15% to 30% moisture. Excess moisture can be eliminated and aeration fostered by adding some (5%) dry straw, leaves or manure. The further addition of 1% to 2% soil or fine sand will enhance the functioning of the worm digestive systems. A solid under floor must be provided to prevent the worms from escaping into the ground.

Worms grown in manure beds are excellent sources of protein for poultry, pigs and/or fish, or in some cases they are eaten by man. Worms reduce the ammonia and other odours, the flies, and the faecal mass.

ANIMALS PROVIDE POWER

Some animals can provide man with power to carry his burdens, pull his carts and tillage equipment, to thresh his grain, operate his irrigation pumps and provide transportation. It is estimated that over three-quarters of all agricultural power in the world comes from animals. Using his own labour a man must expend about 400 hours of heavy toil to turn a hectare of land for planting. One hour of ox-power can substitute for 3 to 5 of these man-hours.

ANIMALS SERVE AS STORAGE

Animals may serve as a storage reservoir of energy, protein and other nutrients that are produced in grain, forage and other crops. The plant growth in excess of current needs is eaten by the animal and converted into animal tissue (growth and fat) that can later be consumed by man.

ANIMALS PROVIDE MEDICINE

Medicinal and other products can be derived from animals: e.g. enough insulin can be extracted from the pancreases of 26 steers to keep one diabetic alive for one year.

ANIMALS PROVIDE COMPANIONSHIP

Small animals keep young children occupied while parents work in the fields. The human soul finds contentment in associating with animals. This can be a disadvantage, however, since once an animal becomes a pet there is some reluctance to eat it.

ANIMALS SERVE AS AN ECONOMIC RESOURCE

Being an easily expandable property asset, animals are valuable trading tools that can usually be readily sold to meet cash needs or emergencies. Maintaining a diversity of animals provides a hedge against disaster or disease. Animals may improve the economic status of the family by providing such products as live animals, eggs, milk, meat or pelts for sale.

Small animals are best adapted to enterprises with limited resources.

Those best suited to the small-scale agriculture scheme include chickens, rabbits and goats because:

1. Small animal inclusion is less likely to upset the crop balance established for meeting other basic family needs.
2. Small animals individually cost less and are therefore more affordable to small-scale farmers.
3. The low individual cost and the short time to get animals into production reduce both the risk of loss and the time needed for investment return.
4. The feed requirements for an individual small animal are minimal and fit well into the limited resources of small-scale farmers.
5. Small animals individually yield small quantities of animal products. The yields of eggs, milk or carcass meat more nearly match the daily needs of the family and will not require preservation.
6. Children and women can be involved totally in the management and care of small animals.

CONSTRAINTS

The constraints involved in the introduction of small animals onto farms with limited resources are:

1. Recognition that small animals are more or less completely dependent upon the family's care, and the animals require more buildings and facilities and a higher level of management skills than do grazing animals. Profitable animal production depends on protecting them from harsh climatic conditions and caging them for control, protection, sanitation and disease prevention.
2. Since small animals are especially susceptible to predators such as dogs, snakes and birds, cages and fences must be provided for safekeeping. In some areas animals may also need protection from theft.
3. An adequate supply of feed throughout the year is required. Cropping systems must be available that will meet the requirements for animal feed while not depriving the family of necessary food or cash crops. This process must include harvesting and storing procedures to maximise the feeding value of the crop. Where feed supplies are limited there is often a tendency to keep more animals than can be adequately fed. It is only when an animal eats more than it needs for maintenance that it is able to produce.
4. A programme for improved health and well-being of the animals is needed. A programme must be adopted that will prevent and/or control the major animal diseases and parasites of the area. This would include quarantine, sanitation, vaccination and breaking the cycle of internal parasites through the caging of chickens and rabbits to keep them off the ground (Figure 1). This caging, together with the tethering of goats, would permit closer supervision and control of the animals.
5. Animals of an improved hereditary makeup (genotype) are not always readily available. Among the native (criollo) animals desirable traits

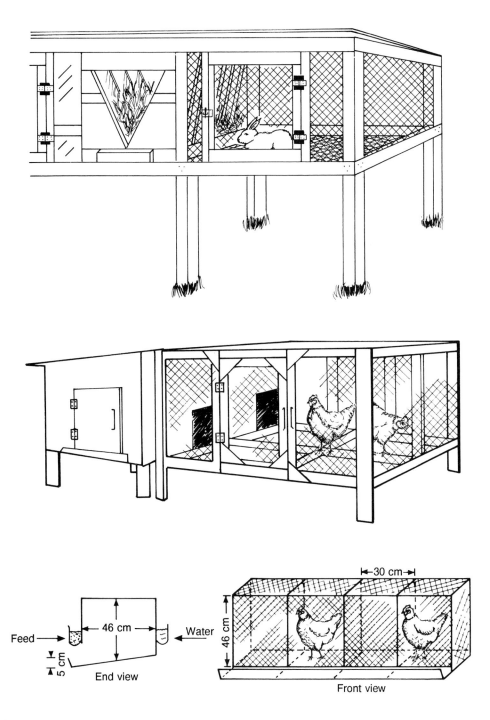

Feed →

← 46 cm →

← Water

5 cm

End view

←30 cm→

46 cm

Front view

Figure 1   Rabbit hutches and cages for broilers or laying hens should be simple, inexpensive, convenient, but must protect the flock.

that can be improved by selection and breeding should be identified. Ways must also be developed for introducing any genetically superior productive characteristics of exotic breeds.

6. A market for animal products produced in excess of family needs is often inaccessible or lacking.

7. Within the families of small-scale farmers, there is frequently a reluctance to accept new technologies and to change traditional practices and be willing to invest capital and labour into improvements basic to the effective production of food from animals.

Of all the factors to be overcome in the production of small animals for food, poor management of proper nutrition and the control of diseases and parasites is the most important.

Animals play an important role in the well-being of the human race. The proper selection and care of animals, coordinated with appropriate crops, can contribute significantly in raising the quality of life while fulfilling the dream of self-sufficiency.

ANIMAL SELECTION

The animals to be selected will depend upon family needs and expectations. It is necessary that the animal species chosen by the family will be able to thrive under the environmental conditions provided, including the feed produced on the farm. The family must decide the kind and amount of the various animal products they want, then choose the livestock that can best meet their expectations. Where animals are not a part of the traditional value system of the family, the introduction of animals will require adjustments in their lives.

A mix of animals suitable for many areas is:

1. 12 laying hens to provide 2800 eggs per year and 11 kg (24 lb) of meat when birds are replaced.
2. 24 broilers replaced every 2 months to produce 150 kg (330 lb) of meat from 144 birds, or 4 does and a buck rabbit to produce 220 kg (480 lb) of meat yearly.
3. 2 does and a buck goat to produce 912 litres (1600 pints) of milk and 16 kg (35 lb) of meat per year.

Other species such as guinea pigs, sheep, pigs, turkeys, ducks, pigeons and/or fish might be used depending on local conditions and choices.

2

THE PARTS AND FUNCTIONS OF ANIMALS

The animal body is made up of a number of systems that, when acting in a coordinated manner, account for what we call purposeful living. To detect the abnormal or diseased animal, it is necessary to know the appearance and behaviour of the normal. Knowledge of these systems is important for many reasons, including:

selecting and preparing feed ingredients,
treating animals that have injuries,
assisting animals in giving birth to their young (parturition),
testing for pregnancy,
recognising symptoms of nutritional and infectious diseases,
administering medicines,
selecting individual animals,
training, mating, milking,
slaughtering animals and cutting, chilling and preserving the meat,
harvesting and preserving animal products,
and many other management practices.

Furthermore, this knowledge provides a basis for enlightened and innovative animal husbandry.

Systems with Little Management Intervention

Except where they are involved with disease, the following body systems do not ordinarily enter into the everyday decisions of management.

SKELETAL SYSTEM

The skeletal system consists of a variable number of bones, cartilages and ligaments arranged in order to:

1. Provide protection to vital organs (brain, spinal cord, heart, lungs, urogenital system).
2. Give rigidity and form to the body.
3. Provide a system of levers for the muscular system, allowing for locomotion, defence, offence, grasping, eating, mating, etc.
4. Help maintain a constant chemical balance within the body. Minerals such as calcium, phosphorus and magnesium can be stored in bone in times of plenty to be drawn upon when demand for the minerals is greater than that obtained from the diet, as in pregnancy and lactation.

5. Provide a site for blood formation. Red blood cells are produced in the marrow and spongy bone substance.

Bones are living tissues with blood and lymphatic vessels and nerves; they are subject to disease, can repair themselves and adapt to conditions of stress. Organic matter makes up about 30% of bone and gives resilience and toughness, while the inorganic salts (largely calcium and phosphorus) make up 45% of the bone and give hardness and rigidity; 99% of the calcium and 75% of the phosphorus of the body are found in the bones and teeth.

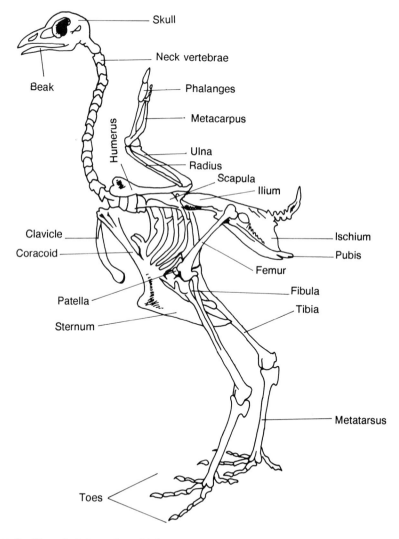

Figure 2 The skeleton of a chicken.

Some of the differences between birds and mammals can be seen in the sketches of the skeletons of a chicken and that of the goat shown in Figures 2 and 3.

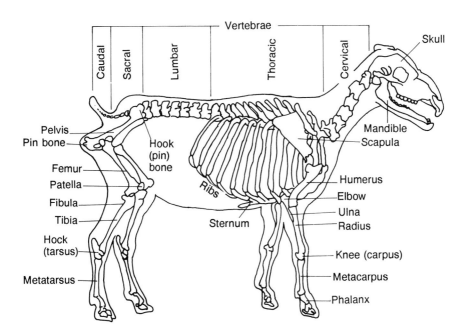

Figure 3   The skeleton of a goat.

MUSCULAR SYSTEM

Muscle tissue is made up of fibres (cells) that have one special property—the ability to contract when stimulated and then later relax. Muscle fibres either contract to their maximum, or they do not contract at all (the 'all or none' principle). Factors causing a response (stimuli) might be nervous, chemical, electrical or mechanical in nature. Energy from carbohydrates, fats and proteins is converted through a series of biochemical reactions to the compound ATP (adenosine triphosphate). ATP is the basic currency (ultimate source) of energy. The splitting of a phosphate bond provides the ultimate energy for a muscle to contract, for a nerve to transmit an impulse, or for a gland to produce its secretion. Drawing a comparison with a car, ATP might be thought of as the electrical energy produced in the alternator and stored in the battery. In the car, the fuel supplies energy to drive the alternator. This generated energy sparks the ignition of fuel or energises the lamps to provide light or does whatever is needed for the operation of the machine. Heat is produced as a by-product and is generally considered a waste. Food energy is converted into ATP.

This energy is expended when muscles contract to produce movement; heat is a by-product. If the muscle is overworked, lactic acid (a chemical intermediate) accumulates, producing soreness and ultimately cramps (tetany).

Three kinds of muscle tissue exist:

1. Bundles of voluntary (striated) muscle work together to move the bones of the skeleton. Under the coordinating influence of the nervous system, these muscles frequently work in pairs, one relaxing when the other contracts to produce meaningful, efficient movement. These muscles constitute the meaty portion of the body and are sources of food for man.
2. A second type of muscle is involuntary (smooth, unstriated) which is found in systems that mainly function automatically without any conscious control. Such muscles are found in the walls of the digestive tract, urogenital system and blood vessels. Smooth muscles are stimulated by the autonomic nervous system as well as by hormones and certain drugs.
3. Cardiac (heart) is a third type of muscle which is both striated and involuntary. These muscle fibres are arranged in such a way that the heart can control itself as well as being regulated by the autonomic nervous system.

NERVOUS SYSTEM

The body systems are coordinated in their function by the nervous and endocrine systems. Irritability, interpretation and conductivity are the special properties of the nervous system. This allows the animal to adjust to its environment to protect, nourish and perpetuate itself. The nerve cells (neurons) are interconnected to receive and interpret stimuli, then transmit impulses to a part of the body for action. The junction between one nerve cell and another is known as a synapse.

The nervous system is integrated into:

1. Voluntary nerves of the brain, spinal cord and peripheral nerves that emerge from the skull and vertebra.
2. Autonomic (involuntary) nerves of the sympathetic (fight or flight) system and the parasympathetic (relaxation and repose) system.

Behavioural patterns (some inherited and some acquired) are associated with the nervous system. These include: aggressiveness, establishing a social order of dominance (pecking order), gregarious (herding) instinct, protecting, curiosity, eating, drinking, sexual behaviour and shelter seeking. Species differ in behaviour as in sleeping patterns; for example, goats do not close their eyes in sleep as do dogs and cats, but do undergo periods of somnolence characterised by loss of muscular tone, drooping ears, etc.

12

ENDOCRINE SYSTEM

The internal functioning of the body is affected by the endocrine as well as the nervous system. Hormones are chemical compounds produced by the body that regulate the rate at which various biochemical reactions proceed in the body. The hormones are produced in the endocrine glands that are located in various places in the body. Hormones are transferred via the blood to an organ or tissue (target tissue) that is affected.

Genes often determine both the rate at which glands secrete their hormones and the sensitivity with which target tissues respond to the hormonal or the neural stimuli acting on them.

RESPIRATORY SYSTEM

One of the most essential needs of animals is oxygen either from the air or from water, in the case of fish. Although land animals may survive for weeks without food and for days without water, they can survive for only a few minutes without oxygen. The major function of the respiratory system is to supply this oxygen while removing carbon dioxide. In addition, the respiratory system dissipates heat, a process which is very important in hot weather. Also vocalisation is made possible by the respiratory system.

A bellows-like thoracic cavity is formed by the rib cage and an upwardly convexed diaphragm. Ribs are curved and flexibly fixed to the spinal column at the back and the sternum in front. They are pulled (rotated) upward and outward by the intercostal muscles. The muscles of the diaphragm contract, flattening it, thus enlarging the volume of the chest cavity, creating a vacuum into which air is drawn, filling the lungs. The passages which warm and filter the air while conveying it in and out of the lungs are the nostrils, nasal cavity, pharynx, larynx, trachea and bronchi. Air is taken by bronchioles within the spongy lung to the alveoli; an alveolus is the smallest subsection of the lung. This arrangement is much like a bunch of grapes, the stem and branches being the trachea and bronchi and the grapes representing the alveoli. Within the walls of these tiny sacs oxygen passes into and carbon dioxide out of the blood. The blood when oxygenated takes on a bright red colour.

The respiratory system of birds is markedly different from and more efficient than that of mammals. In the bird the diaphragm has no respiratory function. The lungs, which are attached to the thoracic wall, expand and contract but not to the extent they do in mammals. The bronchi, lungs and 9 air sacs are all interconnected. Thoracic and abdominal muscles, the sternum and the ribs change the body volume. Bellows-like, these structures move air in and out of 9 air sacs, the air passing through the lungs. Instead of having sac-like alveoli like mammalian lungs, the bird lung has tubular para bronchi, continuous tubes that allow the air to pass through the lungs in only one direction. Two breathing cycles are required to move air through the system. With the first inspiration, air moves from the outside into the posterior air sacs; the muscles contract, the sternum moves upward reducing the body space and creating the first expiration, which moves the air into the lungs. The second inspiration moves the air into the anterior sacs to be expelled from the body with the second expiratory effort. An intricate mechanism makes this movement of air through the lungs unidirectional, reducing respiratory air shunts as in mammals, thereby increasing the efficiency of ventilation.

The respiration rate of an animal is increased by many things such as exercise, excitement, pain, infection and fever. Hyperventilation from breathing deeply and rapidly reduces carbon dioxide, thus causing alkalosis and producing dizziness. Overventilation in animals that are panting in hot weather is avoided to some extent by shallow breathing, moving air through the air passages only; this minimises the alveolar exchange but allows for the cooling effect of water evaporation from the surfaces of the passageways. Hyperventilation is a factor responsible for the poor shell quality of eggs laid in hot weather. Egg shell is composed of calcium carbonate and panting reduces the amount of carbon dioxide available to synthesise the carbonate.

CIRCULATORY SYSTEM

The functions of the circulatory system are to:

1. Make possible the gaseous exchange in the lungs by transporting the oxygen from the lungs to the body tissues and the carbon dioxide from the tissues to the lungs.
2. Carry nutrients absorbed from the digestive tract to the body tissues, and the metabolic waste products to the kidneys and liver for excretion.
3. Control and equalise body temperature by transferring heat between the deeper structures of the body and the surface.
4. Carry hormones from endocrine glands to target tissues.
5. Assist in water balance. Water is ingested as drinking water and in food and is also produced metabolically in the oxidation of foodstuffs. (Metabolic water accounts for 5% to 10% of the total water needs in farm animals and up to 100% in small desert rodents.) Water is lost from the body in urine, faeces, respiratory evaporation, and from the skin and saliva.
6. Help maintain a neutral reaction (pH) in tissues and body fluids by means of blood buffers such as sodium bicarbonate.
7. Reduce danger from haemorrhage by forming a clot to prevent excess blood loss in case of injuries. Fibrinogen, calcium, prothrombin and an activating agent are necessary for a clot to form.
8. Participate in the formation and distribution of body defence (immunological) factors. White blood cells (leukocytes) and antibodies carry on this defensive function.

The circulatory system consists of a heart and blood and lymph vessels. In birds and mammals a 4-chambered heart pumps the blood under pressure through blood vessels (arteries) to the lungs, brain, muscles, viscera and extremities (Figure 4). The arteries branch into even smaller vessels. Finally, the blood passes through tiny capillaries within the

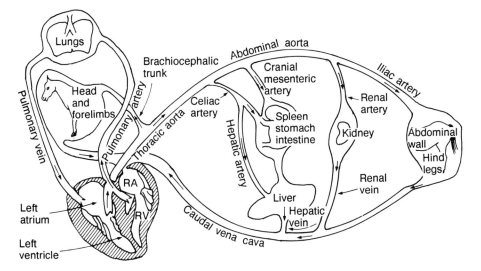

Figure 4   Scheme of blood circulation in the adult.

tissues where nutrient and gaseous exchanges occur; then the blood is returned to the heart through a system of veins. Lymph is a clear watery liquid that leaks from the capillaries and bathes the individual tissue cells. It collects and flows through vessels (lymphatics) to be filtered through strategically placed lymph nodes as it is returned to the heart. Fluid is forced through the veins and lymphatics by bordering skeletal muscle contractions and a series of one-way valves.

The frequency with which the heart beats is called the pulse rate. Smaller animals have faster pulse rates than larger animals. Exercise and excitement will increase the heart beat.

Blood accounts for about 8% of the body weight and consists of:

1. Plasma—the fluid component (90% water, 10% solids). Fibrinogen is the clotting component and principal solid of the plasma. The fluid that remains after the blood has clotted is called serum and contains many elements including the antibodies for immunity and resistance to disease, hormones, nutritive substances, carbon dioxide and minerals.
2. Cellular components—these include (1) the red blood cells which transport oxygen and (2) the blood platelets. Platelets induce the clotting of fibrinogen and also the constriction of blood vessels at the site of injury. (3) The leukocytes (white blood cells) are involved in protecting the body from invading pathogens and other foreign material.

3
SYSTEMS CRITICAL TO DAILY OPERATIONS

The urogenital and digestive systems are especially involved in daily management decisions and practices.

UROGENITAL SYSTEM

The urinary and genital systems are distinctly different in function but since they both utilise some of the same body structures they are often considered together.

1. The urinary system consists of a pair of kidneys in the back part of the abdominal cavity in front of the first few lumbar vertebrae. The kidneys filter water and waste products from the blood, then selectively re-absorb water and some nutrients from the filtrate. By this means the kidney plays a very important role in maintaining a rather constant condition (homeostasis) of the internal environment of the body. Two tubes called ureters transport the urine to the expandable bladder where it is stored until voided through the urethra.
2. The genital (reproductive) system provides the means for producing new individuals to perpetuate the species. There are many similarities between the female and the male and between various mammals and birds, yet there are enough differences to justify their separate consideration. The reproductive tract of the doe goat is sketched in partial cross-section in Figure 5.

The paired gonads in either sex have a dual function: to produce the sex hormones and the reproductive cells (gametes).

Female Reproductive System

The female gonad, the ovary, undergoes cyclical changes; the length of these oestral (heat) cycles and the ways they are affected by the seasons vary among the different species. Seasonal breeders such as sheep and goats are animals that will mate only at certain times of the year. It is primarily the changes in intensity and duration of light (as detected through the eyes) that trigger internal changes in a part of the brain (hypothalamus) in both male and female to bring about the final consequence, mating.

One of the cells which develops near the surface of the ovary is destined to become the egg (ovum). It is surrounded by other cells to form the follicle. This blister-like follicle produces the sex hormone oestrogen which is largely responsible for the growth and maturation of

16

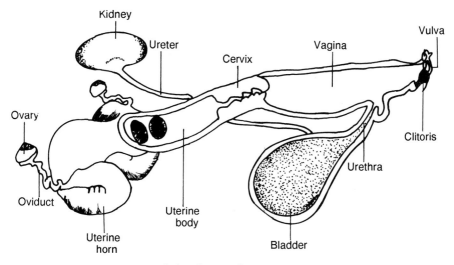

Figure 5 **Urogenital tract of the doe goat.**

the structures involved in reproduction (uterus, mammary gland, etc.). Oestrogen is also responsible for the development of the secondary sexual characteristics with which we associate the qualities of femininity. Oestrogen, with some progesterone, also produces the behaviour pattern known as heat (oestrus), which is receptiveness to the male. When the follicle has matured sufficiently, it ruptures and discharges the ovum into the oviduct (fallopian tube). The ruptured ovarian follicle then changes into the corpus luteum (yellow body) and begins producing progesterone, a second sex hormone which, as the name suggests, is the hormone necessary for pregnancy. Progesterone affects the inner lining of the uterus so that the embryo can be nourished and attached to the uterine wall. Progesterone also fosters the development of the milk secreting cells (alveoli) in the mammary gland.

Following mating (either naturally or artificially), the sperm work upward in the female reproductive tract partly by means of the whip-like action of their tails but primarily by the peristaltic action of the uterus and oviducts. Fertilisation (conception) occurs within the oviduct before the fertilised ovum (zygote) passes into the uterus where it implants itself into the uterine wall to grow and develop (gestate) until ready for birth as a new individual. The relative location of the pregnant uterus is depicted in Figure 6.

A fibro-muscular structure, the cervix, forms a mouth for the uterus, closing it and holding the developing offspring. The cervix remains closed except at oestrus when it allows the sperm to enter and at parturition when the conceptus (foetus) is born.

This new individual in the uterus is called an embryo for the first trimester. During this time differentiation of new tissues occurs; that is, cells that will ultimately become the liver become different from those that

17

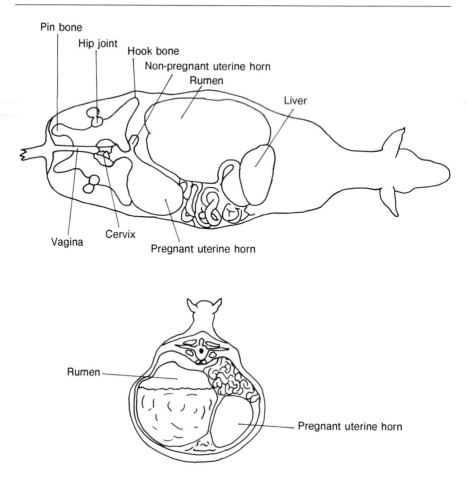

Pin bone
Hip joint
Hook bone
Non-pregnant uterine horn
Rumen
Liver
Vagina
Cervix
Pregnant uterine horn
Rumen
Pregnant uterine horn

Figure 6   Pregnant uterus and other abdominal organs in the goat.
Schematic depiction of the contents of the abdominal and pelvic cavities to
show the location of the rumen and pregnant uterine horn.

will become the eye, or the leg, etc. For the balance of gestation, the new
individual is called a foetus. It develops and surrounds itself with 3 membranes which become the afterbirth (placenta). At the time of birth (parturition), the pelvic ligaments, vagina and vulva relax, the cervix expands,
and the smooth muscles of the uterus and abdominal muscles contract
under the influence of the hormone oxytocin from the posterior pituitary
gland. This forces the foetus through the birth canal (vagina) and the
exterior vulva. In normal birth, both front feet appear first with the head
nestled between them. Usually the young will be born without outside
help. However, assistance will be needed if the offspring is excessively
large or is in an abnormal position. If the dam has strained for an hour

or more and no part of the young has appeared, giving assistance is justified. Within a few hours postpartum (but before 3 days) the afterbirth will normally be discharged. Among the factors responsible for retained placentas are malnutrition, especially a deficiency of vitamin A, disease such as brucellosis, premature birth and heat stress.

Lactation

The mammary gland is perceived as a modified sweat gland; it develops and functions under the influence of the hormones related to reproduction. Figure 7 depicts the internal structures of the udder. The teats and supporting structures of the gland develop under the influence of oestrogen. It is progesterone, however, that stimulates the development of the functional alveoli which secrete the milk. In some species the hormone prolactin from the anterior pituitary gland is primarily responsible for initiating milk secretion after parturition and then maintaining milk secretion throughout the lactation period. Nursing and other stimuli induce the posterior pituitary gland to produce oxytocin, which then causes the smooth muscles surrounding the alveoli to contract. This puts pressure on the contents of the alveoli, forcing the milk into the gland ducts and cisterns. It is now ready to be removed by suckling or milking. Lactation substantially increases the nutrient requirement. Not only do the raw materials going into the milk have to be provided, but a great deal of metabolic work is required to convert them into milk. In the process, approximately 400 kg of blood must be pumped through the udder for each kg of milk secreted. However, lactation improves the efficiency with which energy is metabolised.

The alveoli of the active gland are constantly secreting milk into the lumen of the alveolus. However, at least in ruminants, if the milk is not removed after about 6 hours back pressure develops. Milk secretion diminishes and ceases when this back pressure becomes more than one-quarter the blood pressure. This explains the need for regular milking and why emptying the gland more frequently (up to 4 times daily) results in higher milk production.

Milking only once daily or at irregular intervals drastically reduces milk yield. Twice daily milking is the general rule. For heavily lactating females, milking 3 times daily can increase yield by 15% and an additional 5% increase can follow 4 times milking.

The animal should be comfortable and at ease while being milked. If the lactating female is disturbed or is uncomfortable at milking, the adrenal glands produce epinephrine (adrenalin) which blocks the oxytocin effect, inhibiting milk let-down and removal.

Milk is the only food item found in nature that is prepared for the sole purpose of nutrition. Since this is the material on which the newborn animal survives and grows, it would be expected to be a most nearly perfect food for man also. A few individuals develop a sensitivity to the protein of milk and develop anaphylactic (allergic) reactions when milk is consumed. Other individuals and races (especially orientals and blacks)

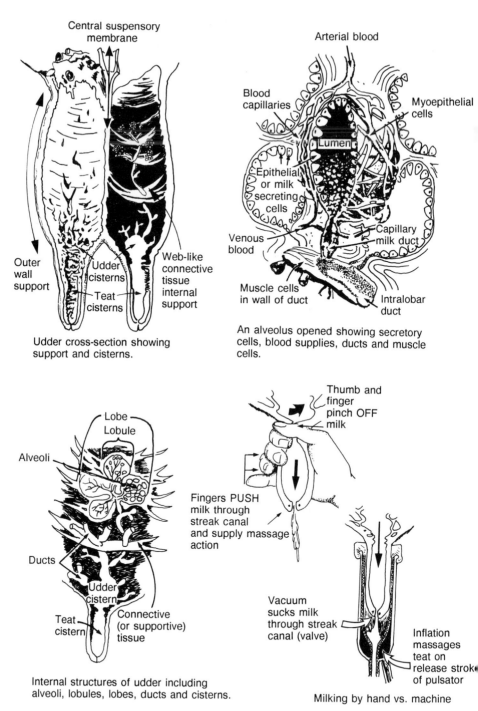

Central suspensory membrane

Outer wall support

Udder cisterns

Teat cisterns

Web-like connective tissue internal support

Udder cross-section showing support and cisterns.

Arterial blood

Blood capillaries

Myoepithelial cells

Lumen

Epithelial or milk secreting cells

Venous blood

Capillary milk duct

Muscle cells in wall of duct

Intralobar duct

An alveolus opened showing secretory cells, blood supplies, ducts and muscle cells.

Lobe

Lobule

Alveoli

Ducts

Udder cistern

Teat cistern

Connective (or supportive) tissue

Internal structures of udder including alveoli, lobules, lobes, ducts and cisterns.

Thumb and finger pinch OFF milk

Fingers PUSH milk through streak canal and supply massage action

Vacuum sucks milk through streak canal (valve)

Inflation massages teat on release stroke of pulsator

Milking by hand vs. machine

Figure 7 Udder structures and milking principles. *(Drawings courtesy of Babson Bros Co., Naperville, Illinois.)*

20

after weaning fail to continue producing the enzyme lactase; then, instead of being digested and absorbed in the small intestine, the milk sugar passes into the large intestine. Here, the intestinal lining is irritated and gas-forming organisms ferment the sugar, producing varying levels of intestinal distress. Those people at risk need to exercise care in the consumption of most unfermented dairy products.

The composition of milk is determined largely by genetics and varies with the species. Milk content is related to the rate of growth of the newborn. Nature did not intend that milk should remain the only food indefinitely. On a solely milk diet anaemia develops as a consequence of milk's deficiency in iron, copper and magnesium; and there is not enough energy to sustain rapid body growth.

As a general rule, milk is remarkably constant in composition. Ordinarily, the yield is decreased before a milk with an unusual composition is produced. It will always have the same osmotic pressure. Mastitis causes a rise in salt (chlorides) at the expense of milk sugar (lactose). The fat content is depressed by factors such as heat, stress and feeding high grain, low fibre, or ground (finely divided) diets. Milk fat content is increased by feeding high roughage diets and saturated fat or whole oil seeds like cottonseed or soybean.

Approximate values for the composition of milk in percentage units are:

	Protein	Fat	Lactose	Mineral	Water
Goat	4	4	5	0.8	87
Rabbit	13	9	1	2.2	74
Human	2	4.7	7	0.3	87
Cow	3.5	3.8	5	0.7	87

The mammary system has been so highly developed in some species and breeds, such as the dairy cow and dairy goat, that the udder has become highly susceptible to mismanagement and infection (mastitis).

Male Reproductive System

The male reproductive organs consist of 2 testicles (testes) suspended in the scrotum, sperm ducts, accessory sex glands and a delivery system (Figure 8).

Each testis produces billions of spermatozoa (germ cells). These are tiny tadpole-like structures that develop within many tiny tubules (seminiferous tubules). They pass from many into a single convoluted tube (epididymis). As the sperm continue to move outward, they pass into a larger tube (vas deferens) which empties into the urethra. During this time they mature, so that when ejaculated with the secretions of accessory sex glands they become capacitated and can fertilise the egg (ovum). The fluid semen contains both the sperm and secretions of accessory sex glands.

The accessory sex glands consist of the paired seminal vesicles, a single prostate and paired bulbo-urethral glands. The seminal vesicles and the

21

prostate produce a seminal fluid that dilutes, carries and nourishes the sperm. The bulbo-urethral (Cowper's) glands produce a fluid that cleanses the urethra before ejaculation.

The urethra is a long tube extending from the urinary bladder through the penis to the exterior carrying both urine and semen.

The penis is the male copulatory organ which deposits the semen within the female reproductive tract.

In order for the testes to produce viable sperm they must be maintained at a relatively constant temperature slightly cooler than the body. For this reason they are located outside the body in a muscular sac (the scrotum). The muscles of the scrotum, when sensitised by testosterone, will contract in cooler weather and relax in warmer weather. This action maintains a constant temperature for the testes by drawing them closer to or moving them away from the warm body.

In mating, the cavernous tissue of the penis becomes engorged with venous blood under pressure. This stiffens the penis so that it can enter the female vagina to deposit the sperm. The sperm are then transported by their own movement aided materially by the contractions of the uterus

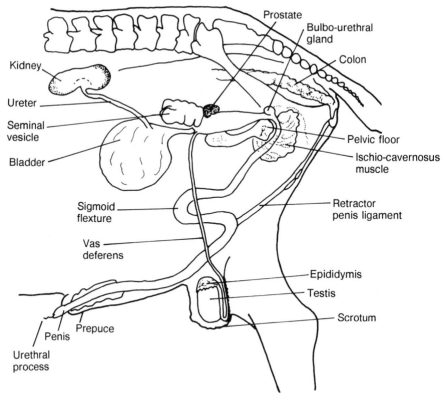

Figure 8   *Urogenital tract of the buck goat.*

to the oviducts where one enters the ovum and joins the male genetic material with that of the female (conception) in order to produce a new individual.

The male hormone (testosterone) is produced in special cells (interstitial cells) located among the seminiferous tubules. Testosterone stimulates the development of the secondary sex characteristics such as deep voice, thicker, heavier bone structure, aggressiveness and other physical characteristics and behaviour patterns that are associated with masculinity. The hormone (but not viable sperm) is produced even if the testes remain within the body cavity (cryptorchidism).

The Urogenital System of a Bird

The urogenital tract of a bird differs in several ways from that of the mammal. The kidneys are situated along the fused backbone. Kidneys produce a cream coloured urine high in uric acid. Urine is transported through the ureters to the cloaca from which it is voided as a part of faeces. The cloaca collects the end products of the digestive, urinary and reproductive systems.

Since birds gestate within a shell they must be very conservative in the use of water; thus they do not have the luxury of producing a watery urine. The end product of their degraded proteins is poorly soluble uric acid, whereas mammals produce the much more soluble urea.

In the female bird only the left ovary, oviduct and uterus are functional (Figure 9). At hatching the chick has 3500 to 4000 small ova. These ova later form a germinal disc that can be seen on the top side of the yolk in a broken egg. The yolk is fully developed within the ovary. The oviduct is divided into 3 rather definite regions: the funnel (infundibulum) that catches the ovulated ovum with attached yolk, the magnum, the area in which the albumin or egg white is secreted and the isthmus where shell membranes are formed. The uterus lays down the shell and the completed egg passes through the short vagina just before it is expelled through the cloaca and vent. In the chicken about 26 hours are required for the process from ovulation to laying (oviposition).

The male bird's reproductive system consists simply of 2 testes located along the backbone inside the body cavity. Each testis has an epididymis and vas deferens that empties through a papilla in the cloaca. A rudimentary copulatory organ is located between these two papillae. Figure 10 shows these structures together with the kidneys and ureters. In mating the male treads (mounts) the female, and the cloaca of both birds touch (cloacal kiss) as the semen is ejaculated. In the female the sperm are then stored in crypt-like glands near the utero-vaginal junction. Prior to ovulation some of these sperm are released and carried by peristaltic contractions to the upper end of the oviduct where fertilisation occurs. Sperm remain viable in this system for 3 weeks or longer, fertilising the eggs laid during that time.

Fortunately, fertilisation is not necessary for egg laying. In fact, infertile eggs stay fresh longer than those that have been fertilised, and they are just as nutritious when eaten.

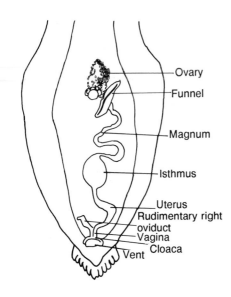

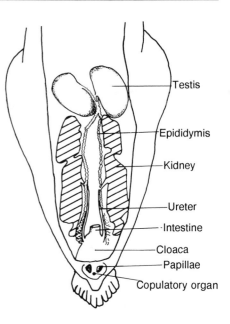

Figure 9   Reproductive tract of the hen. Only the left side is functional.

Figure 10   Urogenital tract of the cockerel.

Artificial Insemination

Artificial insemination has proved of great value in some species and geographical areas. Perhaps the greatest use and success is in dairy cattle in the developed areas. Techniques have been developed and commercial applications have also been made with goats, turkeys, chickens, rabbits and horses. Artificial insemination permits the genetic base to be improved through the more widespread use of outstanding sires; artificial insemination reduces the number of male animals that are needed, thus allowing greater selection pressure. The control of some diseases is helped by artificial insemination. Using an artificial vagina, semen is collected from the male, diluted and stored fresh or frozen until it is inseminated into the female at the proper stage of oestrus. Less of a poorer quality semen can be collected by other means such as electrical stimulation. Cleanliness and attention to detail are essential to a successful artificial insemination programme.

General Considerations

1. The time of initial breeding for an animal is influenced as much by body size as by age. Animals generally reach puberty at about one-third mature weight.
2. The heat period (oestrus) of a female can be readily detected in most species by changes in behaviour and appearance of the genitalia.

24

3. In each oestral cycle there is a best time to mate for optimal fertilisation. The time is usually in the latter part of heat.
4. In most classes of livestock, feeding so that the animals are gaining weight at breeding time, a practice called 'flushing', will improve reproductive efficiency. In animals in which multiple births are characteristic, flushing increases the numbers of ova produced with subsequent increased numbers of offspring. The extra feed sends a physiological message that there will be adequate nutrients for the offspring. This stimulates the natural animal instinct to reproduce at the maximum. Efficient reproduction is dependent on adequate nutrition.
5. Animals generally do not need assistance in giving birth to their young (parturition).
6. The environment should be adjusted to ensure the survival of the newborn. This requires giving attention to temperature and freedom from draughts and dampness. Mammals require colostrum to provide immunity to many diseases. The effect of colostrum is most pronounced in the first 24 hours and becomes nonexistent after 3 days postpartum. An adequate, balanced diet is necessary for healthy growth. A sanitation programme is necessary for freedom from disease. It is easier to adjust the environment to meet the needs of the animal than it is to change the animal genetically to fit the environment.

DIGESTIVE SYSTEM

Good nutrition is necessary for maintenance, growth, production and reproduction as well as health and well-being. The digestive system takes on special economic significance because feed accounts for the greatest cost of production, as much as 90% in some cases.

Organs of Digestion and Their Functions

The body depends on the digestive system to perform several functions. The digestive tract must supply not only the chemical building materials for growth and repair, but also the compounds to supply the energy needed to assemble and utilise these building elements. The structural components of the digestive tract of a rabbit can serve as a reference (Figure 11). The functions these parts serve are:

1. The mouth with its lips, cheeks, tongue, palate and teeth takes food into the digestive tract, chews, moistens and swallows it.
2. The pharynx or crossroads, properly directs food away from the trachea into the oesophagus, a fibro-muscular tube which then transports food to the stomach.
3. The stomach stores food from one meal to the next. The stomach also secretes enzymes to begin digesting protein and, in carnivores, fat. In herbivores, the stomach, caecum and other structures are modified to serve as fermentation vats in which complex carbohydrates such as cellulose and hemicellulose are anaerobically fermented. The end products of this fermentation include volatile fatty acids (VFA) and

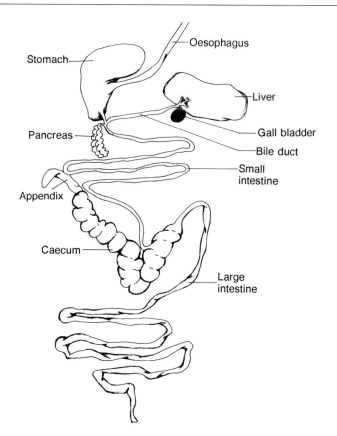

Figure 11 Digestive tract of a rabbit.

microbial cells. By metabolising the VFA and the digested microbial cells the animal can thrive on rough, fibrous feeds. To carry on this fermentative function, the stomach of ruminants has been modified into four compartments: rumen (paunch), reticulum (honeycomb), omasum (many-plies) and abomasum (true stomach) (Figure 12). This fermentation occurs primarily in the caecum and large intestine of non-ruminant herbivores.

4. The small intestine is the primary organ of digestion and absorption. It mixes the secretions of the pancreas and the bile from the liver with its own enzymes for the digestion of carbohydrates, fats and proteins. It then absorbs the end products of digestion, which are simple sugars, free fatty acids and glycerol and amino acids.

5. The caecum is a blind pouch into which the ingesta from the small intestine pours. Anaerobic fermentation and by-product absorption occur here. Some animals (like the chicken) have two caeca.

6. The large intestine (colon) receives undigested residue from the caecum. Some additional fermentation and absorption of fluids occurs in the large intestine.
7. The rectum holds the undigested residue (faeces) until voided through the anus.

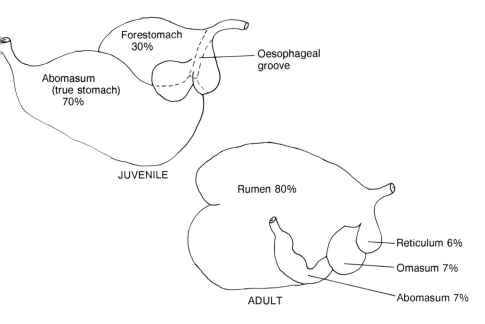

Figure 12   The juvenile and mature ruminant stomach.

The digestive organs of the bird (Figure 13) differ in that the lips and teeth have been replaced by a horny beak; storage occurs in an oesophageal enlargement (the crop). The stomach is modified in that the glandular proventriculus is followed by the addition of a muscular grinding organ, the gizzard. There are two caeca, and the rectum becomes the cloaca, which receives not only the residues of the digestive tract but those of the kidney and reproductive tract as well. The anus is called a vent.

Digestion, Fermentation and Absorption
Before the feedstuffs can enter the body for metabolism they must first be digested. Digestion is the sum total of the mechanical and chemical breakdown of complex feedstuffs into smaller units which allow their absorption into the bloodstream.

Mechanical breakdown   Mechanical fractionation comes from the chewing teeth and the agitation produced by peristaltic contractions of the stomach and intestine. In birds grinding takes place in the gizzard. The smaller feed

27

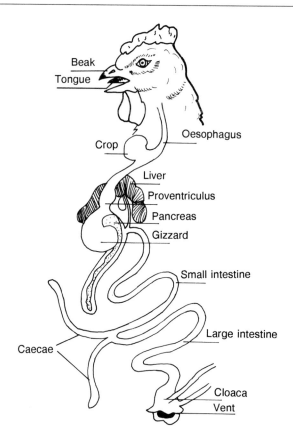

Figure 13 Digestive tract of a chicken.

particles provide a larger surface for the chemical (hydrolytic) action of digestive enzymes. Ruminants, in the process of eating, initially chew the feed only enough to swallow it. The reticulo-rumen constantly churns the ingesta about. The coarser particles are worked toward the oesophageal opening where they are regurgitated, re-chewed and re-swallowed (cud chewing or rumination). The finer, heavier particles are worked through the omasum where fluids are removed before they move on into the abomasum for enzymatic action.

Enzymatic (chemical) breakdown Enzymes are very specific as to the molecular linkages and configurations that they will hydrolyse (split with water). This specificity is much like a lock and key. It takes a special key to fit into and activate a given lock. To prevent digestive enzymes from digesting the tissues that produce them, enzymes are secreted as pro-enzymes.

28

The pro-enzymes are changed into true enzymes only after they enter the lumen of the gut.

It is through enzymatic action that large, complex nutrient molecules are split into simpler ones that can be transferred through the gut wall into the bloodstream. Once in the bloodstream, the nutrients can be transported to the body tissues where they can be used. Thus starch (a complex molecule) is enzymatically digested (hydrolysed—split with water) into glucose, the proteins into amino acids, and the fats into fatty acids and glycerol so that they can be absorbed. Bile from the liver helps digest and absorb fat. The digestive enzymes are produced principally in the stomach, pancreas and small intestine. Saliva from omnivores (including man) contains a starch-splitting enzyme.

Fermentation Animals have no enzymes capable of hydrolysing some complex carbohydrates such as fibrous cellulose and hemicellulose of plants. To take advantage of the energy in such compounds, plant-eating animals (herbivores) have modified their digestive systems to include a place for fermentation. These are organs that hold fibrous feeds for several hours/days and which are stocked with microorganisms (bacteria, protozoa and anaerobic fungi) that can split the chemical bonds of plant fibres. However, without oxygen the microbes can degrade the fibre only to the volatile fatty acids (principally, acetic, propionic and butyric). Oxygen (available to the host but not the microbe) is required to carry the breakdown further to carbon dioxide (CO_2) and water (H_2O).

Speaking of the formation of volatile fatty acids (VFA), it is common knowledge that fruit juices will turn into vinegar when their sugars (carbohydrates) are fermented by the right kind of microorganisms. The same principle holds when fibre (carbohydrate) is fermented by microorganisms in the digestive tract. In this case, though, being much more complex, not only is acetic acid (vinegar) produced but other volatile fatty acids (principally propionic and butyric) as well. Different rations eaten by the animal alter the ratio in which these volatile fatty acids are produced. Acetic and propionic acids are of special interest: roughages foster the growth of microorganisms that produce more acetic acid, whereas grains foster propionic acid formation. This altered ratio, in turn, affects the metabolism of the animal. The more the animal depends on the VFAs for its energy source, the greater the effect of altering the VFA ratio; ruminants are especially susceptible. Milk and milk fat production is increased with more acetic acid and less propionic acid, whereas body fattening is stimulated by more propionic acid and less acetic acid, a circumstance characteristic of high grain feeding for animal fattening. The most efficient use of feed energy occurs when grain makes up about one half of the total ration. The efficiency with which fattening occurs increases until grain makes up about 85% of the ration. Feeding grain beyond this point leads to the formation and accumulation of lactic acid that causes acidosis, anorexia, an erosion of the lining of the rumen and 'founder' or laminitis. In the rumen the short-chain or volatile fatty acids are normally neutralised by sodium bicarbonate and other buffers from the saliva and are absorbed into the bloodstream in the same proportions in which they are produced.

The VFAs thus formed contain about 85% of the energy which was present in the original plant material. Not only do the microbes make this energy available to the host animal, but many of the microbes themselves are digested. In the development of its cell body, the microbe synthesises all the factors necessary for life including essential amino acids and water soluble vitamins and vitamin K. Then, when these microbial cells are digested, these essential nutrient factors are released to be absorbed by the animal. Not all, but many of these microbes are digested. This process, to a varying extent, frees the host from a need for dietary B vitamins and specific amino acids. Ruminants are at an advantage over other herbivores because this fermentation takes place at the beginning of the digestive tract which exposes the fermentive residues to the action of all the digestive enzymes. Thus, ruminants especially are efficient in converting roughages with low quality protein into highly palatable and nutritious milk and meat which contain balanced protein and B-complex vitamins. They can also make use of non-protein nitrogen. The rumen microbes make effective use of simple nitrogen sources (e.g. urea, ammonium salts) to combine with carbohydrates to synthesise protein for their cellular bodies which, as indicated above, become available to the ruminant when these microbes are digested. By observing some precautions, up to one-half of the total digestible protein needed by the ruminant can be supplied as non-protein nitrogen.

Component Separation and Recycling

Some herbivores, such as the rabbit, have developed a different strategy for handling vegetative matter. First, they are very selective, eating only the more digestible parts of the plant. Second, in the large intestine they quickly separate the digesta into digestible and non-digestible portions. By means of **coprophagy** (consuming a soft 'caecal pellet' as it emerges from the anus) they recycle the more digestible parts for further fermentation and digestive action, and void the bulky, indigestible fibrous portions in a hard round faecal pellet. This frees the animal from the necessity of maintaining a bulky fermentation vat, which other herbivores find necessary in order to utilise fibrous plant material. Rabbits are thus able to benefit from digesting some of the caecal and large colon microbes and meet a fraction of their vitamin and amino acid needs. Chickens and pigs are similarly benefited by coprophagy, but to a lesser extent.

Absorption   Absorption is the passing of a substance from the gut into the bloodstream. Some substances (e.g. some drugs and pesticides) can be absorbed through the skin. Water and certain other substances can be absorbed from the body cavity and/or from several body organs. Nutrients from the food are absorbed from the digestive tract. The end products of digestion are simple sugars, amino acids, fatty acids and glycerol. These are absorbed from the small intestine. Water is absorbed primarily from the large intestine. The products of fermentation (volatile fatty acids) are absorbed primarily from the site of their formation, whether it be the rumen, caecum or large intestine.

Absorption generally occurs as a result of: (a) diffusion, in which molecules passively move from a place of their higher concentration to a region of lower concentration, and (b) active transport, which accounts for the most absorption, requiring the input of energy to move the nutrients through the membranes of the gastro-intestinal tract.

Species Differences

Undigested feed particles pass through the digestive system of omnivores in about 1 day whereas 3 to 4 days are required for feed to pass through ruminants. The differences in functions of the digestive tracts among animals are quantitative (a matter of degree) not qualitative. With a few exceptions, the same enzymes are produced and the same digestive and fermentative processes occur in all animals, including man, but some of these processes are relatively more important in one species than in another. For example, VFAs supply the major part of the energy for ruminants but have only minor significance for man. Because of the location of its fermentation vat, the ruminant has a distinct advantage in digesting roughages, but this advantage is lost when low fibre diets are consumed.

The apparent digestibility of a nutrient is the difference between the amount of the nutrient in the feed and that in the faeces and can be expressed as a percentage. Digestibility is not constant for either a feed or a species. Some typical values of digestibility are seen in Table 1.

Table 1 Representative Values of Digestibility for Different Species (%)

	Dry Matter	Crude Protein	Crude Fat	Crude Fibre	NFE*
Lucerne hay					
Goat	59	74	19	41	69
Pig	37	47	14	22	49
Rabbit	39	57	21	14	51
Low fibre diet					
Goat	79	76	90	—	89
Pig	91	92	71	—	95
Human	90	89	84	—	94

* NFE = Nitrogen-free-extract, thought to be carbohydrates.

Nutrients

There are 6 classes of nutrients, each of which is essential for well-being and life. They are provided naturally in the feedstuffs. Thus, animals do not eat nutrients, they eat feedstuffs which contain nutrients. The nutrients are: carbohydrates, fats (lipids), proteins, vitamins, minerals and water. These are derived from the feedstuffs through digestion and are absorbed from the gut into the bloodstream to be metabolised within the body cells. Energy, although sometimes listed as a nutrient, is nothing tangible, but only a quality provided by carbohydrates, fats and proteins.

Carbohydrates account for about three-quarters of a typical ration. They are the primary source of dietary energy. After the energy needs of the body have been met, any excess is converted to fat and deposited in the body. Carbohydrates consist of a combination of carbon (C), hydrogen (H) and oxygen (O) with hydrogen and oxygen being at or near the same proportions as in water (H_2O). Carbohydrates include the simple and complex sugars that are generally digestible. Glucose, $C_6H_{12}O_6$, is the best known simple sugar. The more complex carbohydrates, called polysaccharides, include the digestible starches and indigestible fibre.

Starch is made up of a multitude of glucose units fastened together by the alpha linkage. The sugar- and starch-splitting enzymes of the body can hydrolyse these linkages, thus freeing the glucose for absorption. Fibre generally comes from the cell walls of plants. These cell walls contain, primarily, cellulose, hemicellulose, pectin and lignin. Cellulose, like starch, consists of several thousand glucose molecules linked in long, unbranched chains. The type of bonding in cellulose is referred to as the beta linkage, an arrangement that resists splitting by animal digestive enzymes. Hemicellulose has similar resistant bonding. Nor does the animal body produce enzymes that will disrupt the bonds holding together the structural units of pectin. Different microorganisms do have the capacity in varying degrees to ferment these fibrous components, making at least a portion of the energy and nutrients in the feed fibre available to the animal.

Lignin is a highly complex undigested component of plant fibre which develops as encrustations on older plant cell walls, thus decreasing their digestibility. Lignin is not a carbohydrate.

Carbohydrates are formed by plants from carbon dioxide in the air and water taken from the soil. The sun provides energy to promote the reaction in a process called photosynthesis. The energy of the sun is thus captured and stored in a form that man and his animals can use to maintain their lives. Free oxygen (O_2) is a fortunate by-product of this reaction. A balance in nature is reached. Plants utilise CO_2 and liberate O_2 which is used by animals which in turn exhale CO_2.

Some species have survived on experimental carbohydrate-free diets, yet carbohydrates make up the greater part of practical diets.

Glucose is an essential constituent of blood, yet the animal has no special dietary requirement for this or any specific feed carbohydrate. The body can synthesise any carbohydrate it might need. This is different from the case of some other nutrients, e.g. vitamins and essential fatty and amino acids which have to be included in the diet.

Fats   The animal body can vary in composition from less than 5% to more than 45% fat depending on conditions. Fats (solid) and oils (liquid) are often called lipids. They are made up of 3 fatty acids attached to a glycerol molecule. They are concentrated sources of energy consisting of carbon and hydrogen with only a few oxygen atoms. Fats have 2.25 times more energy per gram than is found in carbohydrates or proteins. Fat is a form in which the body stores energy that is in excess of its current needs.

Animal feeds contain some fat: wheat has 1.8%, lucerne hay 2.3% and soybeans 17.5%. Fat from animal sources tends to be more saturated and harder than the fat from plant sources. Animals do not thrive on a fat-free diet; 3% to 6% fat is required by all farm animals. There is a specific requirement for one fatty acid, linoleic. Fats are carriers for the fat-soluble vitamins. Also, within the body they serve cushioning and insulating functions. Fats reduce dustiness and can increase palatability of the rations.

Proteins contain not only C, H and O but are the body's source of nitrogen (N) and sulphur (S). Small amounts of phosphorus (P) are also present. The primary function of protein in the diet is to supply the amino acids needed for animal maintenance, growth and reproduction, and to supply nitrogen for general metabolism. The dietary protein in the gut is hydrolysed by proteolytic enzymes into the component amino acids. These amino acids are then absorbed and transported by the blood to the liver, glands and muscle where they are reconstituted under the control of DNA into the specific proteins needed by the animal. Proteins on average are 16% nitrogen; therefore, the protein in a substance can be estimated by determining the nitrogen content and multiplying it by 100/16 or 6.25. This calculated 'crude protein' includes not only N from true protein but also N from non-protein sources such as urea and plant nitrates.

Protein is the principal component of connective tissue including blood, muscle and glandular tissue. Protein is also an important component of bone. Enzymes produced in the digestive tract and other body cells are special proteins that serve important roles in the digestion of feed and in the reactions of life (metabolism). Antibodies are special proteins that combat invading pathogens and toxins. There are 20 to 25 different amino acids which enter into the makeup of proteins; 8 to 11 of these amino acids cannot be formed in the animal body (or cannot be formed fast enough to meet the need) and must be obtained from the feed; they are called essential amino acids. Eight amino acids are required by all animals. These are lysine, methionine (contains sulphur), leucine, isoleucine, valine, phenylalanine, threonine and tryptophan. Two others, histidine and arginine, are also generally needed for growth. These are what are called the 10 essential amino acids. The B vitamins pyridoxine, folic acid and vitamin B_{12} are essential for normal protein formation. Whether or not these substances are needed in the diet will depend on how much of them are synthesised in the rumen, caecum and large intestine and absorbed.

For the animal, the biological value of a dietary protein depends on its ability to supply the amino acids in the same proportions as they are needed by the animal. Because the content of individual amino acids varies among proteins, these proteins will have different nutritional values. Proteins derived from animal sources are generally of higher quality than those from plants. Plant proteins from the pulses (legumes) generally are deficient in the sulphur-bearing amino acid methionine. Lysine as well as methionine is generally deficient in the cereal grains. Proteins are burned for energy when there is not enough energy available from carbohydrates and fats to meet basic energy needs. When the diet supplies more protein than is needed to provide amino acid needs, the body burns the excess amino acids for energy. In these cases, the nitrogen of the protein is excreted in the urine.

Dietary protein must supply nitrogen (N) and specific chemical structures such as in essential amino acids. Proteins are needed not only to maintain life and be deposited as a consequence of body growth, but to replace those proteins lost to the body in eggs and milk; however, physical activity (work) does not measurably increase protein requirements; the same general rules apply to vitamins and minerals. Those minerals lost in sweat are exceptions.

Protein can be considered either as total (crude) protein (N x 6.25) or as digestible protein. Digestible protein is generally the preferred measurement for non-ruminants, but crude protein is as satisfactory as digestible protein when feeding ruminants because of the action in the rumen. In fact, non-protein N such as urea can substitute for some of the true protein in ruminant rations.

The animal has a requirement for a specific quantity (grams) of protein, and its requirements can be stated in such terms. However, protein content is often expressed as a percentage of the total diet.

Vitamins are unrelated to each other chemically, yet they all serve essential roles in metabolism. If they cannot be formed in the gut or the body tissue, they must be supplied in the diet. As with essential amino acids, animals differ in the ability of their body tissues or of the microbes in the digestive system to synthesise them. Green, growing plants, fermentative organisms, body tissues and sunlight are the ultimate sources of the vitamins. Vitamins can be classified as fat-soluble (vitamins A, D, E, K) and water-soluble B-complex vitamins and vitamin C.

The animal must have vitamin A to maintain the integrity of its epithelial membranes and for the process of vision. A deficiency of vitamin A results in such things as night blindness, rough skin, diarrhoea, abortion and pneumonia. Vitamin A, as such, is not found in plants, but its precursors are in most plant pigments. Not all plant pigments give rise to the same vitamin A activity; B-carotene is the most effective. Vitamin A is produced when its precursor is altered by an enzyme. Some animals convert carotene to vitamin A more efficiently than others. Carotene is measured in milligrams and vitamin A activity can be measured in terms of International Units (IU). One IU of vitamin A is equivalent to 1.0 micrograms of B-carotene in the chicken, 1.8 in pigs, 1.8 in man, 2.5 in goats and 3.0 in cattle, reflecting the efficiency with which carotene is converted to vitamin A. The chicken is much more efficient in this conversion than is the cow or goat. More recently, vitamin A has been renamed retinol, and the biological activities of all vitamin active materials should be designated as 'retinol equivalent' and measured in micrograms.

Vitamin D regulates calcium and phosphorus metabolism as in bone formation. A deficiency of vitamin D in young animals results in rickets. Vitamin D is produced when the ultraviolet light of the sun strikes precursors (7-dehydrocholesterol) in the skin of animals or on the surface of harvested plants (ergosterol) converting these precursors into vitamin D_3 and D_2, respectively. Vitamin D_3 is also known as cholecalciferol, and D_2 as ergocalciferol. Vitamin D_2 is converted to D_3 which is then changed to the active metabolite 1,25 dehydrocholecalciferol by enzymes in the liver and kidney. Except for poultry most animals can efficiently convert vitamin D_2 to D_3.

The body uses vitamin E (alpha-tocopherol) as an antioxidant and, with selenium, to protect itself from the harmful effects of some free radicals such as those formed when unsaturated fats are metabolised. An absence of either or both vitamin E and selenium results in a muscular dystrophy (white muscle disease) and in chickens a nervous disorder. Vitamin E is found in the oil of the reproductive part of seeds and in green forages.

Vitamin K (phylloquinone) is required by the liver to produce prothrombin, a factor needed for the blood to clot and thus prevent haemorrhage. Vitamin K is produced by fermenting organisms of the digestive tract, usually in sufficient amounts to meet the needs of the animal. Green plants and oil seeds are other natural sources of vitamin K.

The B vitamins include thiamin (B_1), riboflavin (B_2), niacin and pantothenic acid which are primarily involved in energy metabolism (as indicated below). Pyridoxine (B_6) is required for protein metabolism and biotin must be present for fat to be metabolised. Folacin and vitamin B_{12} (cobalamin) are essential for blood formation, choline provides methyl groups (-CH_3), and vitamin C (ascorbic acid) is required for collagen formation. Only primates (like man), the guinea pig, and certain fruit-eating birds require dietary vitamin C. Other animals have the capacity to form this vitamin in their tissues as it is needed.

Minerals constitute the ash that remains when feedstuffs are burned. They make up 3% to 5% of the animal body. Minerals are important structurally as the hardening factors in bone, serve as components of enzyme systems, and maintain the internal body equilibrium (pH and osmotic pressure). Minerals are classified according to the relative amounts needed by the body as either macrominerals (calcium, phosphorus, magnesium, sodium, potassium, chlorine and sulphur) or micro or trace minerals (iron, zinc, manganese, copper, cobalt [a constituent of vitamin B_{12}], molybdenum, chromium, selenium, silicon, iodine, fluorine).

Some elements like calcium, phosphorus and magnesium perform a structural function in bone. But even these, as do the other minerals, serve in regulating and carrying out the processes of life such as energy metabolism, blood formation, respiration, cellular osmotic pressure and acid-base balance. Examples might be the roles played by iron and copper in haemoglobin formation and oxygen transport, of iodine in thyroxine to regulate the rate of metabolism, and of calcium and sodium in the ability of the muscles and nerves to respond to a stimulus. Phosphorus has more known functions than any other mineral element. It is located in every cell of the body and enters into many metabolic processes, including energy metabolism within the cells.

Vitamins and minerals are usually measured in terms of grams; however, some vitamins are expressed in units of activity (the quantity of the nutrient that will produce a given response when consumed by the animal). The requirements of some macrominerals like calcium, phosphorus and salt can also be expressed as a percentage of the total diet.

Water accounts for up to 95% of the weight of the newborn body and about 75% of the weight of the adult. It serves the body as a solvent to carry oxygen and nutrients to and waste products away from the tissues. It enters into many metabolic reactions, and it functions in maintaining an osmotic pressure balance. Water plays a very important role in regulating the body temperature. First, it can absorb or lose a great deal of heat energy with little change in its own temperature (it takes 1 calorie to change 1 gram of water 1°C). Thus, the fluid medium of the blood can distribute heat from one place in the body to another without a material change in body temperature. Secondly, it has a high latent heat of vaporisation; it absorbs almost 600 calories for each gram of water evaporated. This is important in excess heat dissipating via evaporation from the respiratory system and the skin. There is a constant H_2O loss from the body through urine, faeces, respiration and perspiration. Some desert-adapted animals, such as the camel, donkey, sheep and goat, can lose up to one-fourth of their body weight through dehydration and then regain it again in one drinking.

Thirst is the best indicator of the need for water. Daily intake varies with temperature, humidity, diet, health and activity. Water intake is generally 3 to 8 times the total feed dry matter intake. Water restriction reduces feed intake. A reduced rate of passage through the gut reduces the moisture content of faeces. Milk production increases the water demand of lactating animals. Laying hens have an increased water demand to replace that lost in the egg. The water loss and subsequent requirement is increased by diarrhoea and fever.

Deficiencies of energy and of the nutrients reduce activity, retard growth, delay puberty, reduce fertility and depress milk or egg production. Deficiencies also reduce resistance to infectious diseases and parasites. These deficiencies may result from lack of sufficient intake or, in some cases, may result from drastic nutrient imbalance and from interference with digestion and absorption.

Energy

Energy, as such, is not a nutrient but is a quality possessed by carbohydrates, fats and proteins. Dietary energy can be partitioned into different categories depending on the channels through which it is lost to the body (Figure 14). The value of each of these categories will vary with the species, size, activity and physiological condition of the animal and with the type and nutrient balance of the ration.

A

Gross energy—that energy released when the feedstuff is totally burned.

↓

Digestible energy   +   Faecal energy

↓

Metabolisable energy   +   Urinary and gaseous (CH₄) energy

↓

Net energy   +   Heat increment (HI)

↓

Net energy for   +   Net energy for production
maintenance   (energy in the eggs, milk
(energy for basal   or body weight gain, or
metabolism,   movement and heat
maintaining body   generated from muscular
temperature and   contraction in work)
activity at
maintenance)

B

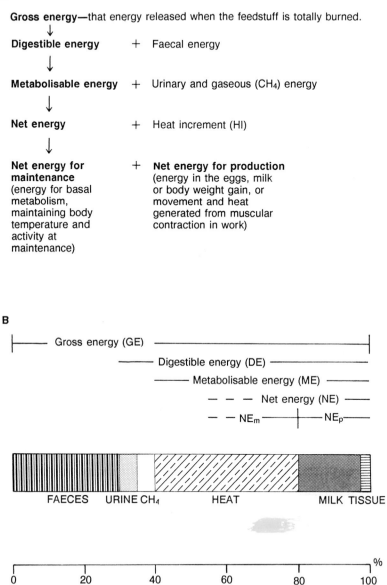

Figure 14   *Schematics illustrating the partition of feed (gross) energy in the animal.*

37

Energy is encountered in nature in many forms such as radiant, electrical, mechanical, chemical and heat. Animal husbandry is concerned primarily with the latter 3 forms. Except for thermoregulation, the form in which energy is utilised is not as heat, but as chemical bonds. For this reason some nutritionists prefer to use the joule as a unit of measurement for energy; this is because the joule can be used to measure mechanical and electrical as well as heat energy whereas the calorie measures only heat energy. The calorie will be used here because it is suitable, more traditional and perhaps more easily understood. The calorie is defined as the amount of heat required to increase the temperature of 1 g of water by 1°C. Calories can be converted to joules by multiplying by 4.1855. To avoid using large numbers in discussing animal requirements the kilo (thousand) and mega (million) calorie are used.

Free energy, 'the fire of life', is the power needed for exercise, food digestion, enzyme and hormone synthesis, milk production, egg laying, bone and muscle development in body growth and maintenance, fat formation and deposition, muscular work, nerve impulse transmission, and any other activity or chemical reaction within the body. Free energy is released when a material is converted into a product that has a lower energy level.

Animals get their energy from the oxidation of the organic compounds, such as carbohydrates, fats and proteins. Oxidation is the chemical process in which oxygen from the air is combined with the carbon and hydrogen in the nutrients to produce carbon dioxide (CO_2) and water (H_2O) with a concurrent release of energy. This oxidation of hydrogen yields much more energy than does the oxidation of carbon. Fat has more H in relation to C than does carbohydrate which partially explains why fat yields more energy than carbohydrate. Oxidation occurs rapidly in a fire, all the energy being released as heat; this reaction for carbohydrate is shown by the formula:

$$1 \text{ g } C_6H_{12}O_6 + 6 O_2 \rightarrow 6 CO_2 + 6 H_2O + 3.75 \text{ kcal of energy}$$

or

1 mol of glucose (180 g) produces 675 kcal of heat

The animal body cannot tolerate this much heat at one time. It has devised a rather complicated set of reactions to release the energy in a step-by-step fashion. An analogy of the nature of this process might be that one can get from the top floor of a building by jumping from a window, or one could descend step-by-step down the stairway. In both cases, one gets to the ground, but the means and consequences are a bit different.

Much of the energy released in these series of reactions is captured in 'high energy' bonds. One mol of glucose (180 g) produces a net of 36 high energy bonds. This represents an efficiency of about 39%; i.e. 61% of the energy in glucose is lost as heat. Thus, it is possible to salvage in these high energy bonds only a fraction of the energy that is released when chemical bonds are pulled apart. The remainder of the energy is lost as heat which is a waste to the body except in cold weather.

Important roles in the breakdown of energy nutrients are played by thiamin and pantothenic acid. These B vitamins are components of coenzymes involved in the metabolic pathways converting the complex molecules of carbohydrates and fats into a simple, 2-carbon acetyl compound similar to acetic acid.

The B vitamins niacin and riboflavin and the minerals iron and copper play central roles as coenzymes in the oxidation of acetyl to carbon dioxide and water while salvaging in high energy bonds the energy that is not lost as heat.

The minerals phosphorus and magnesium play important roles in forming the high energy bonds and in their utilisation in energy metabolism. ATP (adenosine triphosphate) has 3 phosphate radicals attached to an adenosine molecule. Two of these phosphates (the second and third) form high energy bonds. The splitting of this third phosphate bond provides the ultimate energy for all reactions of life, whether for a muscle to contract, for a nerve to transmit an impulse, for a gland to produce its secretion, etc. Biological reactions that require energy to proceed are driven by the splitting of this terminal high energy phosphate bond in ATP yielding adenosine diphosphate (ADP) and an inorganic phosphate plus available free energy. Thus, this high energy bond of ATP becomes the 'currency' of all energy conversions.

Energy is defined as the ability to do work. It is measured and expressed as the amount of heat produced as a result of the animal being alive and working and/or the amount of heat which a feedstuff, body excretion or animal product yields when burned. The calorie, or in many areas the joule (0.23892 cal), is the unit generally used to express energy values.

Energy is of the greatest concern in feeding livestock for several reasons. It is needed not only to support maintenance, growth, egg production, lactation and work, but it is the nutrient most likely to be limiting in most practical feeding situations. If not enough energy-bearing nutrients are available from the feed, the animal will meet its energy needs from stored glycogen and fat or body protein. If the energy needs are met from a variety of complete, natural feedstuffs, requirements for all other nutrients will also generally be met. The energy needed to put on a unit of gain increases with the body weight and/or age of the animal. This is because a unit of gain in a fatter, larger or older animal contains a greater proportion of fat; and fat requires more energy to deposit than does muscle and bone. Energy is used more efficiently for maintenance than for production. However, if an animal is lactating, it will use energy more efficiently for all purposes, even for fattening. Schematics in Figure 14 show how the animal's use of energy is partitioned and the terms used. There are various systems of measurement to express the energy need of an animal and the energy in feedstuffs for meeting that need.

Gross energy is the total amount of energy within the feedstuff. **Digestible energy** results when the energy voided in the faeces is deducted from the gross energy consumed. Further, subtracting the energy in the urine and eructated gases (mostly methane) yields **metabolisable energy. Net energy**

is derived by deducting the heat increment (specific dynamic heat) from metabolisable energy.

Metabolisable energy (ME) is the unit of energy measurement generally used for humans, pigs and poultry. In non-herbivorous animals the heat increment factor is not as great and is rather constant because roughage makes up only a minor part of the diet. Metabolisable energy can be approximated by multiplying the digestible energy by 0.82. The metabolisable energy (physiological fuel) values for carbohydrates, fats and proteins are 4, 9 and 4 kilocalories per gram, respectively.

Heat increment (HI) is the heat produced within an animal as a consequence of eating an increment of a feed. This includes that heat formed in ingesting, masticating, digesting, fermenting, absorbing and metabolising the feed. This is comparable to the heat developed in an engine in the process of converting the chemical energy of the fuel into the mechanical energy in the crankshaft. This heat is wasted energy in hot weather but is profitably used in cold weather to keep the animal warm. Heat increment varies greatly in roughages, sometimes amounting to more than half of the total energy present in the feed; grains have a much lower and more consistent HI than do roughages.

Net energy   Of all the systems for energy measurement, probably the most accurate in moderate to hot climates is the net energy (NE) system, in which losses in the faeces, urine, combustible gas and heat increment are all taken into account. Net energy is that energy which is usable by the body and is partitioned into two components:

1. Net energy for maintenance includes that used for keeping the resting body alive when it is in a thermoneutral environment (basal metabolism), energy needed to move about and do the things animals do when they are neither gaining nor losing weight, and the energy needed to maintain the normal body temperature in either hot or cold environments;
2. Net energy for production is the energy which is actually deposited in the body as tissue growth, or which is in the eggs, milk or wool produced or in the heat and mechanical energy which are produced when muscles contract in work or other movement.

Total digestible nutrients (TDN) is perhaps a simpler, but in a warm climate a less accurate, system of evaluating the energy needs of the animal and the energy content of feedstuffs. Its use is traditional among many in the USA, but it has no advantage over digestible energy (DE). TDN is based on the proximate analysis and digestion trials. Digestibility of a nutrient is considered to be the difference between the amount of the nutrient consumed and that excreted in the faeces. The TDN content of a feed is the total of:

digestible fat × 2.25,
digestible protein,
digestible fibre,
and digestible nitrogen-free-extract (NFE).

The major criticism of TDN as a basis for energy calculations is that it overestimates roughage energy values in comparison to concentrates. This is because roughages have a higher heat increment loss, and TDN does not take heat increment loss into account. However, in cold weather TDN measures the actual or practical value of a feed better than net energy because the heat produced as a consequence of eating the feed (heat increment) is not wasted but is salvaged in keeping the animal warm.

One kg of TDN is equivalent to 4400 kcal DE. A general relationship between NE and TDN is expressed in the formula for estimated net energy (ENE):

$$\text{ENE (mcal/kg)} = (0.0245 \times \text{TDN \%}) - 0.12$$

Note that as the TDN increases, the fibre content declines as would happen in going from a roughage to a concentrate; also, the relative value of the constant (-0.12) in the formula decreases.

There are several other systems of feed evaluation and of expressing energy requirements that are used throughout the world. The energy system used to formulate a diet will depend somewhat on the purpose for which the energy is used as well as the system customarily used in the area. Some of the factors commonly used in evaluation and in converting one method to another are as follows:

% TDN = 0.95% digestible organic matter (DOM)
% DOM = 97.4 (1.63 × % crude fibre in dry matter)

1 kg DOM	= 1.05 kg TDN
1 kg starch equivalent	= 5.082 mcal DE 1.15 kg TDN 1.1 kg DOM
1 kg TDN	= 4.4 mcal DE
% TDN	= 22.68 mcal DE/kg
mcal estimated NE/100 lb DM	= 1.393% TDN − 34.63
mcal DE/kg DM	= 0.0441 × % TDN in DM
1 lb TDN	= 2000 kcal DE in concentrates or 1500 kcal DE in roughages
1 kg Scandinavian feed unit	= 3.44 mcal DE
1 cal ME	= 1.22 cal DE

1 cal DE	= 0.82 cal MF
1 joule (j)	= 0.239 cal
1 calorie	= 4.184 j

The energy requirement of animals depends on the physiological uses to which that energy is put. The maintenance requirement for any nutrient is that needed to keep the animal alive without either gaining or losing weight. The daily requirement for energy consists of:

1. That energy required for basal metabolism (maintenance of life when quiet, not eating, and in a thermoneutral environment). In terms of metabolisable energy, the basal metabolism of all animals requires 70 kcal for each kg of body weight raised to the three-quarter power (metabolic weight).

   70 kcal ME/(kg body wt)$^{0.75}$

2. That energy required for activity at maintenance (getting up, standing, eating, digesting and absorbing the feed, walking to the water trough, fighting the flies, etc.)
3. That energy needed to keep the animal warm or cool (thermoregulation).

The requirement for body growth and fattening is considered to be 3500 kcal ME/lb or 9.5 kcal DE/g. The gross energy deposited in growth is in both the non-fat organic (mostly protein) portion (5.6 kcal/g) and the fat (9.4 kcal/g). The weight increase of thin, non-lactating adults contains 5% to 12% protein and 50% to 75% fat. This represents about 6.5 kcal/g. The increase in body weight will consist of a higher proportion of fat as the animal becomes older and larger. Since fat contains more energy than other tissue, this means that it will require more energy to put on a unit of gain for older animals or those that are larger for their age.

As the animal becomes larger as it matures and ages, the body composition changes; the fat percentage increases at the expense of water and, to a lesser extent, protein. In lambs it was found that for each percentage increase in fat there was a 0.73% decrease in water and only a 0.03% drop in protein.

The relationship of water, fat, protein and ash in the body composition as the animal (pig) grows and matures is illustrated in Figure 15. Both the amount (kg) of these factors in the eviscerated body and their proportion in a unit of gain are pictured. Fat and water have an inverse relationship. While the total amount of water found in the animal body increases as the animal grows larger, this increase of water is at a slower rate; fat shows the opposite phenomenon. Or, stated another way, as the animal becomes larger the proportion of water in each unit of gain declines, whereas the proportion of fat increases. It should be recognised, also, that in young animals at the same body weight, the next unit of gain by faster gaining animals will have the same fat content as the same unit of gain made by a slower growing animal. In growing animals it is the comparative body weight and not the age or rate of gain, *per se*, which influences the percentage of fat in the gain.

The requirement for pregnancy is insignificant during the first trimester.

However, it is during the last trimester that the greatest foetal growth takes place. A feed adjustment of up to double the maintenance requirement needs to be made.

The energy requirement for lactation will depend on the fat and other organic solids in milk. Since the solids are correlated, usually goat and cow milks are converted to a 4% fat-corrected milk (FCM) basis using the formula:

kg FCM = 0.4 (kg milk) + 15 (kg fat)

Milk from the cow and goat contains about 9% non-fat solids. There are about 700 kcal of energy in 1 kg FCM. Production requirements are 1246 kcal ME/kg FCM with 16.28 kcal ME added or subtracted for each 0.5% change in fat. Due to its higher protein and fat content, rabbit milk contains more than twice the energy of goat milk.

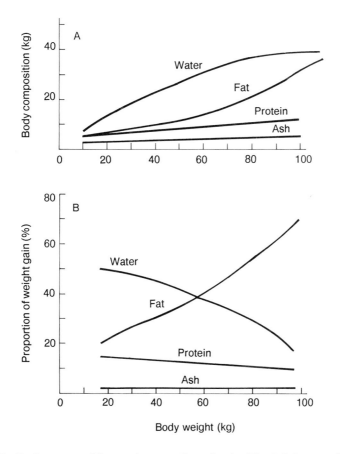

Figure 15 **Body composition and proportion of gain. The total amount (kg) and proportion (%) of water, fat, protein and ash in the eviscerated body of a pig.**

Table 2   Composition of Some Common Feeds (as fed)

Feedstuff	Dry Matter %	Energy Digestible kcal/kg	Energy Metabolisable kcal/kg	Energy ENE	Crude Protein %	Crude Fibre %	Calcium %	Phosphorus %	Carotene mg/kg
Roughages and Forages[1]									
Bermuda grass hay, s-c	91	1960	—	720	8.5	26.0	0.37	0.19	53
Maize silage (dough)	26	820	—	300	2.1	6.5	0.09	0.07	12
Maize stover, s-c	85	2030	—	600	5.7	28.8	0.51	0.08	2
Grass hay, s-c	90	1910	—	880	9.1	28.2	0.51	0.19	44
Kikuyugrass mature	34	880	—	450	2.3	10.6	0.10	0.33	—
Kitchen scraps	12	400	—	—	1.9	1.1	—	—	—
Lucerne—fresh	24	620	510	280	4.9	6.5	0.45	0.06	47
Hay s-c early bloom	89	2430	1020	960	17.7	24.9	1.33	0.23	126
Hay s-c full bloom	88	2280	960	880	13.3	30.6	1.13	0.20	65
Meal (17% protein)	92	2400	1370	1000	17.5	24.4	1.33	0.24	150
Oat straw	91	2000	—	550	4.0	37.0	0.23	0.06	—
Rice hulls	92	440	—	50	3.1	44.5	0.09	0.08	—
Rice straw	91	1800	—	440	4.0	32.0	—	—	—
Soybean straw	88	760	—	410	5.2	44.0	1.59	0.06	—
Energy Feeds									
Barley grain	89	3330	2460	1550	12.4	5.6	0.04	0.33	2
Cassava tubers	37	1310	1190	370	1.3	3.0	—	0.04	—
Maize grain, yellow	89	3500	3380	1760	9.3	4.6	0.03	0.28	4

Molasses, cane	75	2520	1980	1100	3.9	4.5	0.79	0.08	—
Oats grain	89	2950	2550	1430	12.1	10.6	0.06	0.33	—
Quinoa	90	—	—	—	13.0	—	0.02	0.80	—
Sorghum grain	90	3330	3270	1710	11.4	2.4	0.03	0.30	1
Wheat bran	89	2610	1240	1250	15.1	10.3	0.11	1.26	2
Wheat grain	88	3450	2940	1760	14.9	2.9	0.03	0.38	2
Protein Supplements									
Chickpea seeds	90	3460	2100	1760	19.4	6.6	0.17	0.38	—
Cottonseed meal (41%)	90	3090	2240	1580	40.7	12.6	0.17	1.09	—
Milk, skimmed, dried	94	3790	2550	1930	33.6	0.3	1.28	1.02	—
Peanut meal (solv ex)	92	3580	2690	1550	49.9	10.5	0.20	0.63	—
Soybean meal (44%)	91	3270	2450	1709	40.3	6.3	0.29	0.64	—
Soybean seeds, whole	91	4000	3200	1900	39.3	5.4	0.25	0.60	1
Sunflower seeds, whole	94	3300	2790	1760	18.6	25.0	0.17	0.52	—
Mineral Supplements									
Dicalcium phosphate	97	—	—	—	—	—	23.0	18.27	—
Limestone (ground)	99	—	—	—	—	—	33.7	0.02	—
Oyster shell	99	—	—	—	—	—	37.6	0.07	—

s-c = sun cured.   solve ex = solvent extracted.
Metabolisable Energy = 0.82 (Digestible Energy).
ENE = Estimated Net Energy for ruminants.
DE of feedstuffs = 4.41 kcal/g TDN, or 220 kcal/lb TDN.
1 mg carotene = 400 IU vitamin A.
[1] Nutrient value varies with the stage of maturity.

Feedstuffs (Animal Feeds)

Feedstuffs provide the nutrients essential for life. Feedstuffs must be taken into the body and broken down into smaller physical and chemical units so the nutrients can be absorbed into the bloodstream. In practical feeding situations feedstuffs of like composition and characteristics can often be exchanged; e.g. wheat can replace corn or clover can replace lucerne. There are adequate amounts of most nutrients provided in the feedstuffs usually eaten by the animals. The amount of those nutrients that need special attention and are usually found in many common feedstuffs are shown in Table 2. It should be emphasised that these values serve as guides only; they will not indicate the exact value for any specific feedstuff. The digestible energy in various feedstuffs is graphically shown in Figure 16. Grasses grown in tropical areas are less digestible, yielding lower digestible energy values. The stover (fodder) and straw of native varieties generally yield more digestible energy than do the stiffer stemmed, higher yielding varieties.

Animals eat more of the feeds that taste good to them and are more digestible. The palatability of the ration is important because adequate feed intake is needed for maintenance and production. Digestibility is important because this permits ingesta to disappear from the tract giving space for additional intake. To a limited extent the animal can draw on its body reserves, but otherwise, **production occurs only when the animal consumes feed above and beyond its maintenance requirement.** Since plant growth patterns vary cyclically with the wet season, unless other provisions are made, livestock gains are largely lost during succeeding dry periods. The liveweight curve follows an undulating pattern requiring much more time for an animal to reach slaughter weight or production than if fed properly from birth. Drought increases the severity of this trend. This circumstance points to the need for adequate and proper feed harvest and storage procedures. How well grain is preserved in storage depends on the dryness of seeds and their protection from weather and vermin. The quality of stored forage depends on:

type of forage (grass, legume, etc.),
stage of maturity when harvested,
leaf retention,
moisture content,
protection from rain and other elements, and
harvesting and storage methods.

Feedstuffs may be classified as roughages or concentrates.

Roughages are the vegetative leaves and stems, sometimes accompanied by seeds, of legumes, grasses and forbes that contain more than 18% crude fibre in their dry matter.

Crude protein and digestible energy of roughages decline with maturity. As a general rule, to obtain the greatest yields of digestible nutrients, roughages should be harvested just before entering the reproductive

46

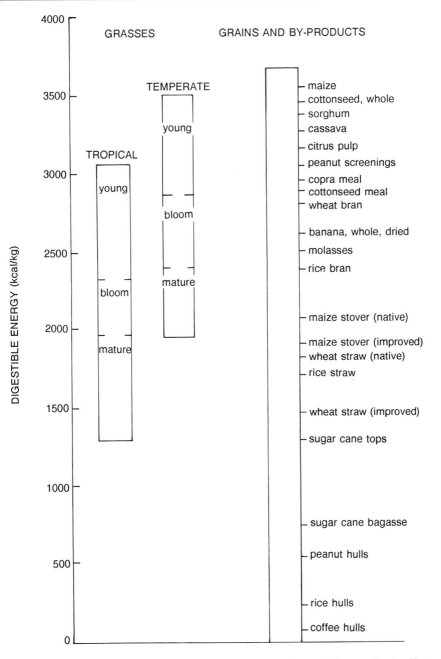

Figure 16 Digestible energy content of grasses, grains and by-products. Note differences in grasses grown in tropical and temperate areas and some native and improved varieties of crops.

stage. Care should be taken to preserve the leaves and to prevent mould development; this goal can be achieved by either reducing the moisture to between 10% and 14% or by excluding the air as in a silo.

Legumes have a higher calcium content and tend to have higher protein than grasses at the same stage of maturity. Grasses have higher crude fibre than legumes; however, because grass fibre contains more cellulose and less lignin, fibre from grasses has a higher digestibility than that from legumes. Plants grown in the tropics have a lower crude protein and higher crude fibre content. Dried mature plants are very low in protein and phosphorus.

Depending on the moisture content and harvesting and storing methods, roughages may be classified as follows:

1. Pasture is forage that is harvested by the animal and directly consumed. Land cannot be farmed as intensively when pastured but often this allows productive use of land unsuitable for tilling. Fencing and/or herding of animals is usually necessary.
2. Soilage (green chop) is fresh forage harvested and brought to the animal for immediate consumption.
3. Tubers and roots—potatoes, beets and cassava are examples.
4. Silage is forage with a moisture content of 60% to 65% that has been packed in an airtight space (silo) and allowed to ferment anaerobically. Some fermentable carbohydrate is necessary for the preserving acids to develop. For this reason it is easier to make good silage with the grain crops than it is with lucerne. Silage-making nutrient losses in the field and from fermentation and seepage are comparatively low, amounting to 15% to 25%. Another advantage of silage is that it can be made in wet weather which would not allow hay making. Furthermore, there is less opportunity for the animals to sort out the more palatable portions and leave the other. Animals thus tend to eat the entire plant, and weedy crops that would make poor quality hay can be ensiled with reasonable success.

   Due to their high moisture content, more pasturage, soilage and silage must be consumed to provide the same amount of dry matter nutrients as in hay. This high moisture might indeed limit the energy intake because the animal may lack enough space in the gastro-intestinal tract to eat sufficient dry matter to meet its needs for maintenance and production.
5. Hay is the harvested plant material which has been dried to less than 14% moisture. Higher moisture can lead to mould development resulting in energy losses from heating. In severe cases, this heating can cause spontaneous combustion. Total nutrient losses from field to animal are generally 25% to 50%.
6. Straw and stover are the dry fibrous residues after separating the heads and ears from grains such as barley and millet.

Concentrates have high nutritional potency because they are so readily digestible. Energy feeds are generally thought of as having less than 18% crude fibre and less than 20% crude protein.

1. Grain (seeds) generally fit into two classes: cereals of maize, wheat, millet, sorghum grain, barley, oats, etc., and pulses such as beans, soybeans, chickpeas, etc.

   Which grain is fed depends on its availability, palatability and cost per unit of energy or nutrient. Cereal grains are essentially devoid of calcium and their protein content and quality varies. Corn (maize) is generally considered the standard by which other grains are compared.

   Compared with cereals, the pulses (seeds of leguminous plants) have a higher protein content, and this protein generally has a higher biological value. However, raw soybeans and other pulses often contain harmful substances (enzymes and/or alkaloids) which can reduce their usefulness unless these factors are destroyed by processes such as heat or extraction. For example, destruction of the trypsin inhibitor in raw soybeans by heat is a function of the temperature, duration of heating, particle size and moisture conditions. Over 95% of the inhibitor can be destroyed by atmospheric steam cooking at 100°C for 15 minutes at 5% moisture. More rigorous conditions are required for other inhibitory factors. No amount of cooking will destroy the solanine that develops in the green skin on potatoes exposed to sunlight.

   The high fat content makes soybeans a rich source of energy as well as supplying a good balance of amino acids. Chickpeas have a protein quality equivalent to soybeans, but have less fat. Whole cottonseed with its high fat, protein and fibre content makes an excellent supplement for ruminants.

   Nature intended that the outside covering on seeds would protect them from damage. This seed hull must be broken up before the seed is available for digestion by the animal. This disruption might be done by crushing, rolling or coarsely grinding the seeds. Although grinding more finely can increase digestibility, such a practice is not justified in view of the increased power costs and the resulting dustiness of the fine texture which makes the feed less acceptable to the animals.

2. By-products include molasses (a palatable liquid which, when mixed up to 10% with other feedstuffs, reduces dustiness and serves as a binder), bran, brewers' grains and oilseed meals (high in protein and phosphorus). Protein supplements contain 20% or more crude protein in their dry matter. The protein supplement of choice is generally soybean meal which is produced when soybeans are ground and the oil is extracted either by heat and pressure or by extracting with a solvent. Soybean meal is not only palatable but is rich in the amino acid lysine which is generally deficient in maize and some other grains. Soybean meal does not contain the trypsin inhibitor found in raw soybeans.

   Cottonseed meal is also a rich protein source, but, while cheaper, its protein lacks the biological value of soybean meal. Cottonseed meal contains varying levels of the toxin gossypol, which limits its use to 5% of the diet of non-ruminants. Iron supplements reduce the ill effects of gossypol.

   Fats and oils are concentrated energy sources which increase energy content, palatability and reduce dustiness and friction in the milling process. Faecal matter is another animal by-product that might be a source of feed. If palatability and disease control problems are resolved, up to 10% of the dry matter and in some cases up to 30% of the total diet can be made up of faecal matter with beneficial results.

Animal by-products such as fish meal and meat meals may be used to supplement cereal grains. Because their protein is of high quality, they are especially valuable in balancing the essential amino acids of a diet. Such items must be treated to prevent transmission of diseases.

3. Supplements include vitamins, minerals and/or synthetic amino acids. Also, certain hormones and antibiotics that may be fed at subtherapeutic levels often get an 8% to 12% positive response in production.

Caution must be exercised that supplements are not misused. Some substances can pass from the feed into meat, eggs or milk and thus should be used with discretion. It is essential that the feeding of antibiotics, hormones and other drugs stop several days before slaughter to allow potentially harmful residues to leave the animal tissues.

Feedstuff Analyses

Proximate Analysis Feedstuffs are evaluated in many ways, one of the most common being the strictly chemical proximate analysis which reports in percentage units:

Moisture—amount of a given feed sample that is evaporated in a drying oven.

Crude protein—total nitrogen × 6.25.

Crude fat (ether extract)—the amount of the feed sample that will dissolve in ether.

Crude fibre—ash-free residue after treating a sample with a weak acid and a weak base.

Ash—mineral residue after burning.

Nitrogen-free-extract (NFE)—a remainder determined by subtracting all the other fractions from 100%. This is a rough indicator of the more digestible carbohydrates but also includes hemicellulose and some of the indigestible lignin. Errors in determining other components are accumulated in the NFE value.

Learning the true digestibility and nutritive value of a feedstuff requires animal trials in which the feed is actually fed and the response measured. The proximate analysis says nothing about how much of the feed components are available to the animal. More sophisticated chemical tests for nutrient analysis and for estimating digestibility are being developed and becoming more popular. These include the Van Soest process, best applied to forages, which separates the easily digested cell contents from the poorly digested cell walls. Neutral detergent fibre consists of the cell walls. Acid detergent will dissolve the hemicellulose leaving the lignocellulose (acid detergent fibre). Sulphuric acid oxidises the cellulose leaving the lignin and the acid-insoluble ash. The ash remains after the lignin is burned. The process has been outlined in Figure 17.

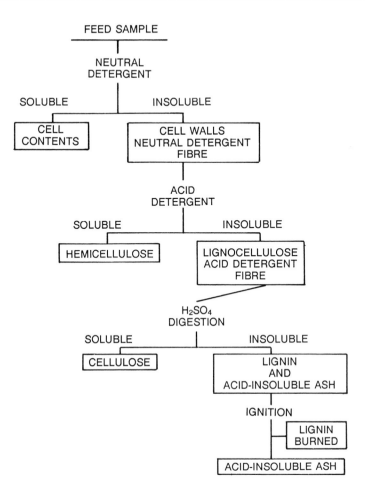

Figure 17   Van Soest method of forage analysis.

Labelling   Commercial feeds will generally be accompanied by a tag or invoice providing information as to the name and address of the manu-facturer, ingredients listed in decreasing order of quantities present, and a guaranteed analysis of the minimum crude protein and crude fat and max-imum crude fibre.

Nutrient Requirements

All of the 6 nutrients are required for life, but the amounts of each vary depending on such things as species and physiological status (age, size, sex, reproductive condition, stress, climatic conditions, etc.). All factors might not be needed in the diet because some of them may be synthesised

51

in the gut or animal tissues. In developing rations to feed livestock, nutrient needs are considered under the categories of maintenance, growth, reproduction (including egg production, gestation and lactation) and work activity.

The requirements, expressed as feeding standards, have been determined for the various classes of livestock by experimentation, practical feeding experience and calculations. Feed composition tables have also been prepared to help determine which and how much feedstuffs to feed to meet the animal's requirement. Both feeding standards and composition tables provide only average values and must be viewed only as guides.

Developing Rations

Animals do not eat nutrients, they consume feedstuffs that contain nutrients. Specific nutrients are required by animals, but there is no requirement for any specific feedstuff. A ration consists of the components and amounts of feedstuffs fed in one day. The husbandman must know the daily requirements of the various nutrients for his animals. Then, taking the feedstuffs that are available, he must evaluate the many nutrients they contain. With this information, these feedstuffs can be combined and processed in the most economical way to meet all of the nutrient requirements of the animal. The feed must be acceptable to the animal so it will eat enough to meet its nutrient needs for production.

Animals eat first of all to meet their energy requirements. Within limits animals have the capability of adjusting their intake to meet energy needs. Until the capacity of the digestive system becomes the limiting factor, the higher the energy requirement, the more an animal eats. Some animals, such as the rabbit, have a greater ability than others to adjust their intake; they eat more of an indigestible feed. When the gut-fill (capacity of the digestive system) is not limiting, the greater the concentration of energy in the diet, the less feed is required to meet energy needs. This is illustrated in Figure 18. Increased fibre reduces the rate of passage in herbivores, except, perhaps, in the rabbit where the more fibrous portions of the feed are separated and eliminated. At levels usually encountered, high fibre reduces the intake of most animals, but tends to increase the feed intake of rabbits. Increased dietary fibre in man and other omnivores tends to increase the rate of passage of ingesta through the gastro-intestinal tract. Other factors that influence feed intake are the amount of glucose and fatty acids in the blood (chemostatic), the need to burn energy to keep warm (thermostatic) and psychological factors such as habituation and competition.

Animals eat more of a more readily digested feed—up to a point. When digestibility exceeds that point, the animal consumes less feed while maintaining the same energy intake. If the energy requirement increases due to lactation, egg production or activity, the feed intake increases correspondingly. If the full capacity of the gastro-intestinal tract has been reached before the energy requirement has been met, the digestibility of the ration should be increased. The total digestible nutrient intake is the product of the total dry matter intake multiplied by the proportion of that dry matter that is digestible.

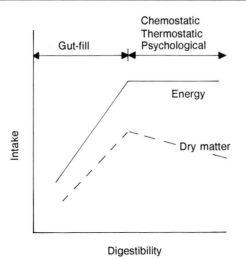

Figure 18   Factors influencing feed intake.

The values for nutrient requirement and feed composition have been determined and published in tabular form in many sources, including animal nutrition and animal husbandry textbooks and government publications. The compositions of a few common feeds have been compiled in Table 2 from these sources. It should be recognised that all these are average values only. Since there are differences in composition of various lots of the same kind of feed and there is variation among animals in their ability to consume and utilise feeds, the feeding standards and composition tables should be used only as general guidelines. The same units of measurement (DE, TDN, g, IU, etc.) must be used for measuring both the requirement and feedstuff composition.

Care must be taken to provide all of the essential nutrients; a deficiency of one will, as the weakest link in a chain, become the limiting factor. And feeding more of any one nutrient than is needed is not only wasteful but in some cases can be dangerous. With few exceptions, if energy requirements are met with a mixture of naturally produced feedstuffs other dietary requirements will likewise be met.

Consideration must be given to the requirements for maintenance, growth, fattening, activity and reproduction, which will vary with different classes of animals and their environment. Besides energy, proteins, fats, vitamins and minerals, other factors are sometimes required, e.g. herbivorous animals demand the presence of a roughage (fibrosity) factor for maintaining the integrity and functioning of the digestive tract. Chickens, unless their feed is ground, will require grit in the diet; grit provides abrasive material for the grinding action in the gizzard.

The Pearson square is a useful tool for determining proportions to combine to obtain a desirable mixture that is intermediate between the two ingredients. Assume that you desired a ration containing 16% protein made up of a mixture of maize (at 10% protein) and soybean meal (at 44% protein).

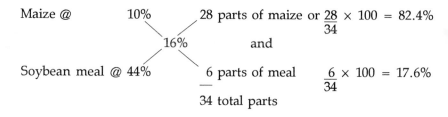

Maize @   10%   28 parts of maize or $\frac{28}{34} \times 100 = 82.4\%$

             16%   and

Soybean meal @ 44%   6 parts of meal $\frac{6}{34} \times 100 = 17.6\%$

            34 total parts

Note that the parts of maize amount to the difference between the percentage protein value for soybean meal and the desired mixture. Similarly the amount of the soybean meal is the difference between the protein in the desired mixture and the maize. The 28 parts maize and 6 parts soybean meal make a mixture of 34 parts. Or, in 100 parts (or %) there would be 82.4 parts (%) maize and 17.6 parts (%) soybean meal, the mixture providing 16% protein.

If 5 kg of a vitamin–mineral mixture that supplied no protein were added, the concentration of protein, and other non-added nutrients as well, would be decreased, having only 95% of the previous levels. In this example, for instance, the protein content would be lowered to 15.18%.

The comparative costs of nutrients can be determined by dividing the percentage of the nutrient in the feedstuff into the unit cost. For example, if maize were 80% TDN and cost £100 per ton and another grain were 77% TDN and cost £98 per ton, the comparative energy cost for maize would be £100/.80 or £125/ton of TDN; for the other grain the cost of TDN would be £98/.77 = £127. Thus, even though the other grain costs less than maize per ton, it would be more economical to supply the energy with maize.

Processing a feedstuff can alter its nutritive value.

1. Grinding increases the digestibility of grains that have a hard seed coat. Fine grinding of either grain or hay is not recommended. This is because fine grinding takes more energy, is more dusty, and it may cause digestive disturbances.
2. Pelleting increases the density of a feed. This decreases the storage space needed and increases the dry matter intake by the animal. The heat involved in the pelleting process sometimes alters the starch and other molecules. Pelleting also prevents sorting by the animal. Animals tend to eat more of a pelleted feed, becoming more productive.
3. Cubing (wafering) of forages increases the density, thus reducing the space required for storage. The pressure generated in the cubing process crushes the stems, reducing the fibrosity of the roughages.
4. Cooking and soaking grains are of benefit only in specialised cases.

It is generally recognised that the faster an animal grows the less expense in feed, time and investment is involved in getting the animal to slaughter or productive size.

Animals become productive only after they consume feed above their maintenance level. Many areas of the world are overstocked; there is only marginally enough feed to maintain the existing animals. Usually there will be insufficient or only enough feed for the adult animals. When young animals are added, with their requirements for growth in addition to maintenance, the total animal productivity declines. This suggests the need to reduce herd numbers by selling or slaughtering for meat those least desirable for breeding. After reducing the size of the flocks or herds, there is usually an increase in total production from the remaining fewer animals. This is because less feed is required for maintenance which leaves a greater proportion of the feed for production.

4
THE ANIMAL IN ITS ENVIRONMENT

MANAGEMENT

Management of the animal might be defined as the process of adjusting the environment to best meet the needs of the animal. At the same time, under practical conditions this exercise must be profitable, which might limit how much is to be done. However, there are necessary aspects to management in addition to providing feed and a disease-free, comfortable environment. These include regularity of appropriate care and consistently doing things that need doing when they need to be done (breeding, feeding, treating diseases, cleaning the watering troughs and feed mangers, keeping adequate records, etc.).

In areas of the world where climatic and other environmental conditions are not optimal, the farmer has several options for caring for his animals. He can:

1. Leave the animals to their own resources in the natural environment. Such animals will give their highest priority to their own comfort and survival. They will eat only enough to meet their own survival needs, which will result in poor production.
2. Introduce improved stock to cross with native stock. Through selection and breeding this process can produce a type of animal with a greater degree of adaption and production capacity.
3. Introduce improved exotic stock and try to increase its resistance to a harsh environment. This is a slow process with limited likelihood of success in doing anything other than increasing its survival ability at the expense of its productive capacities.
4. Change the environment to meet the needs of higher producing animals. Provide needed shelter from the elements; produce and store acceptable feeds; apply sanitary measures and protect from diseases; adopt management measures for the animals' comfort and well-being. Through these means a farmer can achieve the most rapid progress with greatest production increase at minimum costs.

CLIMATE

The animal has an optimal range of climatic factors for its maximum production. For example, the comfort zone for rabbits is 15°C to 18°C (60°F to 65°F) and for chickens 20°C to 30°C. Animals generally are more tolerant of low than of high temperatures. For goats, sheep, cattle, buffalo, pigs and poultry optimal climatic conditions would be something like an air temperature of 13°C to 20°C, a wind velocity of 5 to 18 km/hr, relative humidity of 55% to 65% and a moderate level of sunshine. These factors are interrelated; for example, decreasing the relative humidity would permit the animal to be comfortable in a higher temperature. To maintain the same body temperature (thermohomeostasis) the animal must establish equilibrium between heat gain and heat loss. The heat produced internally together with the heat absorbed from the environment must be balanced with the heat lost through conduction, convection, radiation and evaporation. The herdsman seeks to balance these conditions by providing appropriate shelter and, to a lesser extent, by altering the diet.

Cold Climate

Newborn animals are particularly susceptible to cold. In colder climates animals, especially the young, must be protected from dampness and draughts. Energy requirements and disease susceptibility increase when animals are kept in wet, muddy pens.

In cold weather animals use several stratagems to reduce their heat loss. They hump up and shrink the skin to reduce the body surface area; they reduce the respiration rate. Body heat is preserved by reducing the blood circulation to the periphery such as the skin, legs, ears, etc. Hair erection increases the insulation value of the hair coat. Animals seek protection from moisture and wind. They increase heat production through shivering and eating more feed. Exposed to cold over time, they grow more hair and a layer of fat under the skin

Rain or snow falling on the animal increase the body heat lost by convection and by the evaporation of water trapped on the hair. This loss is increased by wind.

Hot Climate
Animal responses to heat include:

1. Decreasing feed intake to reduce heat increment (the heat produced as a consequence of eating, digesting and utilising feed). Animals with heat resistance more quickly reduce the amount of feed they eat. This may be good for the animal, but it is bad for production. The animal stops eating when its body temperature begins to rise.
2. Seeking shade to protect themselves from the radiant heat of the sun. This advantage of shade is reduced, however, if humidity is high and the shade is limited. If there is not enough shade, animals crowd together to get out of the sun. This crowding then reduces the heat loss from the body surface.
3. Finding cool or wet surfaces on which to lie (wallow).
4. Vasodilation to increase blood flow to the body surface to increase radiation loss.
5. Sweating. Animals that have no sweat glands, like rabbits, obtain some surface cooling from transudation and applying salivary moisture to the hair by licking themselves.
6. Panting and excessive salivation (slobbering) to evaporate water from upper respiratory and oral cavities.

7. Decreased thyroid activity with a subsequently reduced metabolic rate.
8. Increased water intake and subsequent increase in urinary output.
9. Decreased production.
10. Increased body temperature.
11. Increased irritability, coma and death.

When heavily heat stressed, animals usually stand, extend their heads, increase salivation and pant, which reduces heat and even water intake. They do not want to stop panting long enough to eat or drink. Rapid breathing creates a vicious circle because the physical activity of panting itself produces heat. Thus, the more an animal pants, the harder it has to pant to rid itself of the extra heat produced by panting.

Failure to maintain heat balance causes the body temperature to rise. An increase of over 1°C brings negative results. These effects are especially harmful if environmental temperatures are not cool enough during the night to allow the body temperature to return, even temporarily, to normal.

Photoperiod
Another climatic effect on animals is seen in the influence that the length of daylight has on the breeding season of such animals as sheep, goats and buffaloes and many birds. Changes in length of day are registered through the eye and hypothalamus.

Indirect Effects of Climate
The climate not only affects the animal directly, but also indirectly. The quality and quantity of feed, the diseases, the internal and external parasites and the prevailing sanitation problems are all influenced by climate.

CONFINEMENT

Confinement should provide animals with the least stress and most comfort while providing the manager with the most production efficiency and convenience and the least expense. Quarters should be clean, dry and well ventilated, but not draughty. The well-being of the individual animal and the flock or herd should be the major objective of housing and management procedures. Such a circumstance is economically desirable since it both enhances production and minimises losses from morbidity and mortality. The animal's performance is the best indicator of the comfort and well-being of that animal. This is especially true with reproductive traits such as conception, litter size, egg and milk production.

It is easier to handle, feed and treat the animal, to milk or gather the eggs, etc., if the animals are confined in pens or cages. However, their performance can be affected by such matters as the amount of shelter, space on the floor and at the feed and water troughs, presence of other animals and their social relationships (pecking order), type of floor surface, and building materials of pens or cages.

Internal parasites can become more of a problem in tethered and confined animals on the ground. This is because there is a greater possibility of infestation due to a buildup of parasites and their eggs. This risk is greatly reduced if animals are caged off the ground and tethered animals are moved frequently to new ground.

5
DISEASE CONTROL AND SANITATION

Disease control is essential to profitable livestock production. An added concern is that some 200 animal diseases can be transmitted to man. There are many factors relating to disease control, but the most important factors probably are an adequate diet, comfort, isolation and sanitation.

Disease is defined as a malfunctioning or a deviation from the normal functioning of any or all of the tissues and organs of the body. Its effects can be measured by variations in such factors as appetite, alertness and activity, pulse and respiration rate, characteristics of body secretions and excretions, body temperature, pain and/or lesions in various body parts, milk or egg production, body weight growth rate and/or rate of producing offspring. Any departure from the normal function, sensation or appearance generally indicates a disorder or disease. These variations are called symptoms. Many infectious diseases have similar symptoms but are actually caused by different organisms. And sometimes both contagious and non-contagious ailments may produce the same symptoms. This makes identifying the cause of the disease (diagnosis) more difficult. Different symptoms will be observed depending on the body system affected. For example, complications in the respiratory system are seen in heavy breathing, rattle in the throat and a nasal discharge. Ailments in the digestive system might cause anorexia, bloating of the stomach, intestine or caecum, constipation, abdominal pain and tenderness, diarrhoea or abnormal faeces. It is important that the farmer understands the causes of diseases (aetiology) so he can act to prevent their repeated appearance or spread.

INFECTIOUS DISEASES

Animal diseases can result from many causes. **Living organisms** that cause diseases (pathogens) are of many types. Some are communicated to other animals, some are not.

Viruses are so small they can slip through tissues with a minimal disruption and tissue reaction response. Viral diseases are difficult to control because the virus closely resembles the basic genetic material (DNA); consequently, finding a chemical that will kill the virus without killing the body cell itself is difficult. Antibiotics are generally not effective against viruses. The best control is by avoiding the disease and by preventive vaccination. Examples of viral diseases are aftosa (foot and mouth disease), rabies, bluetongue and Newcastle disease.

Rickettsiae and mycoplasmas are organisms intermediate between viruses and bacteria. Some diseases caused by these organisms are: Q fever, chronic respiratory disease (CRD) in poultry, caprin pleuropneumonia and infectious synovitis.

Bacteria are plant-like cells that can be seen only with a microscope. Many bacteria serve useful purposes but others invade and destroy body cells, while others invade body tissues and produce toxins that do damage. Some organisms produce chemicals that destroy other organisms—these chemicals are called antibiotics and are generally useful in treating bacterial diseases. Brucellosis, blackleg, enterotoxaemia, tetanus and tuberculosis are examples of bacterial disease. Anthrax is caused by a spore-forming bacteria.

Moulds and related fungi usually live in the soil and often produce spores that allow propagation after long periods of dormancy. Ringworm and actinomyces (lumpy jaw) are diseases caused by fungi. It is of interest to note that moulds are the usual source of antibiotics.

Parasites include a wide range of organisms from single-celled protozoa to multicellular lice that affect both the inside and outside of the body.

Pathogens and parasites generally enter the body through a body opening such as the mouth, nose, eyes, genitals, navel cord in the newborn, or through a wound. An infected animal sheds pathogens through body excretions and secretions such as faeces, urine, milk, tears, saliva, blood, flesh, placenta or abscess discharges. These pathogens can then be spread from one animal to another by direct physical contact, or indirectly through such things as feed, water, or air-borne moisture or dust, or through some intermediary host such as other animals or insects. The specific diseases to which animals might be exposed will vary from one area to another; usually, information on prevalent diseases, their causes, prevention and treatment is available from governmental or private veterinary agencies.

Whether an animal succumbs to a disease depends on the interaction of several factors:

1. The degree of **animal resistance** may have a genetic origin; some species or strains are naturally less susceptible to some diseases. Resistance may be a quality associated with good health. Nature has provided several mechanisms by which animals protect themselves from disease. The skin and mucous membranes are the first line of defence. The inflammatory reaction is one of the body responses to injury whether infectious, mechanical or chemical. There is an increased blood supply to the injured tissue which brings white blood cells (leukocytes) to fight the invaders and dissolve debris. This extra

61

blood also supplies additional nutrients and carries away wastes. Redness, swelling, and an increased temperature and tenderness are all associated with inflammation. If the reaction is localised, there may be an accumulation of pus (dead cells, leukocytes and the offending microorganisms).

Resistance may result from immunity developed from previous contact with the disease. The antigen–antibody reaction takes place when the body produces an antibody to react specifically with an antigen. Antigens are usually proteins that are foreign to the body; they may be bacteria, viruses or such things as pollen that has passed through the mucous membrane, or undigested protein passing through the intestine into the bloodstream. Parts of the body (the reticulo-endothelial system) are constantly adding to the blood serum complex protein molecules (gamma-globulins). When this system is challenged by an antigen these proteins are altered and become antibodies (modified gamma-globulins) to neutralise the effects of the specific antigen that stimulated their formation. The development of natural immunity is dependent on exposure to pathogens or similar materials.

Resistance to some diseases is enhanced by vaccination. Vaccines are man-made materials used to help the animal develop immunity to specific diseases. Vaccines are the preparations of antigens that produce a reaction leading to antibody formation without actually producing the disease. In subsequent challenges by the pathogen, the antibodies will protect the animal if the antibody level is high enough. The level of an antibody often decreases after a peak response but will increase after re-vaccination or exposure to the disease organism. Some vaccines (bacterins) are killed organisms, some are live, attenuated (weakened) organisms, and some are the attenuated metabolic products (toxoid) that was produced by the pathogen.

2. **Concentration and virulence of the pathogen.** One single microorganism will be unlikely to cause a disease; however, a large number of organisms, especially if virulent, can overwhelm the defence mechanisms of the animal. For this reason, isolated animals or those given ample space are not as prone to develop infections. It is when many animals are concentrated in a limited space, maintained in the same area year after year and/or are closely housed that diseases become rampant.

PARASITIC DISEASES

Intensive management systems, heavily grazed areas, various tethering systems and animals being in the immediate vicinity of heavily used watering places all favour worm infestation. If animals are ill-fed or sick, infestation with internal parasites is more likely and the effects produced by the parasites are more obvious. The parasitic effects may go unrecognised, yet they are insidious and often grow progressively worse. Symptoms include unthriftiness, diarrhoea, loss of weight, pot belly and anaemia.

Animal parasites are of many types and produce varied consequences.

Protozoa are single-celled, microscopic animals some of which cause internal parasitic diseases in animals. Coccidiosis (Eimeria) and anaplasmosis are examples of protozoal diseases.

Worms (helminths), of which there are over 60 species found in birds alone, are internal parasites that involve the digestive tract, lungs and other tissues. Roundworms (nematodes) include ascarides (large roundworm), strongyloides (lungworms and threadworms), capillaria (hairworms) and hookworms. Depending on the type, nematodes can be transmitted directly from animal to animal or have intermediate hosts such as earthworms, grasshoppers, beetles, sowbugs (woodlice) or cockroaches.

Flatworms include tapeworms (cestodes) and have a head and segments. They have as intermediary hosts houseflies, ants, slugs, snails, grasshoppers, beetles, birds, rodents and other wild animals.

Flukes (trematodes) require habitats near water and intermediate hosts of fish, crustaceans or vegetation. Different genera infest the blood, intestine, liver and lung.

Insect larvae (maggots) of certain flies such as the screw fly and warble fly affect various tissues of the body.

External parasites (ectoparasites) include pests such as ticks, mites (mange), and the insects fleas, lice, mosquitos and horn flies. These pests greatly annoy animals as well as suck blood and transmit other diseases.

Sanitary measures of quarantine and cleanliness are designed to reduce the introduction of pathogens and physically and chemically to lessen the number of any damaging organisms present. Such measures should minimise stress on the animal body, recognising that stress makes the body more susceptible to diseases.

METABOLIC DISEASES

Many diseases are due to faulty metabolism; they take the form of such problems as acetonaemia and milk fever in goats, hypoglycaemia in baby pigs, and blue comb in poultry. Acetonaemia (ketosis) is an accumulation of ketone bodies which are produced in the oxidation of fat. Normally, when carbohydrate metabolism is in equilibrium, fats are completely burned, but if the body, for whatever reason, is unable to metabolise carbohydrate fast enough to meet its energy needs, body fat is only partially burned into ketone bodies which can accumulate in harmful numbers.

Stress

Stresses are physical or mental (psychological) disruptive influences that bring about changes within the animal. They may be of a positive challenging nature, but usually stresses are harmful. Stressors can be physical (e.g. injury or inclement weather) or emotional/psychological (e.g. strict confinement or loud noise). One of the most important factors that produces an emotional stress in an animal is a sense of lack of control over its environment. If a stress is continuous, instead of showing the usual behavioural signs of distress such as moving out of the hot sun into the cool shade, the animal might just give up and come to accept the stressor and fail to resist.

Negative aspects of stress include declines in fertility, growth, production, disease resistance and general health. These effects can be minimised by avoiding stressors especially at critical times such as mating, parturition, weaning or transportation.

PARTURIENT DISEASES

In the female reproductive system some problems associated with giving birth to the young are:

Dystocia (abnormal labour) can be suspected if the offspring is not born within the normal gestation and parturition times. A cow should calve within 3 hours after starting labour, a sow should pass one offspring at least every hour, a doe goat should drop her kids 1 to 3 hours after starting labour, and the doe rabbit should kindle in 30 minutes, with individual kits born at intervals of 1 to 5 minutes.

Causes of dystocia include:

Improper presentation of the foetus such as a head or leg drawn back.
Excessive size of the foetus.
Monstrosities.
Pathological causes resulting from diseases.
Abnormal genital tract of the dam such as a narrow pelvic opening or undeveloped vagina.

Treat dystocia by first tying the animal. Observe cleanliness while giving any needed assistance. The hands, vulval area and equipment should be washed with soap and water and rinsed with a disinfectant. Fingernails should be trimmed and cleaned. First, establish a correct presentation of the foetus, then using an obstetric chain or smooth cord attached to the legs, gently pull outward and downward when the dam strains using ratchet-like retention of pressure.

Caesarian section in otherwise dire circumstances may be performed by a veterinary surgeon.

Retained placenta may be due to malnutrition (vitamin A deficiency etc.) or disease (brucellosis etc.). Special attention to the dam should be given if the afterbirth has not been shed after an appropriate time. Drugs and/or gentle, sustained outward pressure is recommended for removal of the unshed placenta. Do not attempt manual removal of a placenta that does not respond readily to positive pressure. The placenta should remain intact and not be torn into pieces. It can remain for several days without serious consequences. Depending on the value of the animal and the availability of funds and veterinary surgeon, specialised assistance should be used.

NUTRITIONAL DISEASES

Some diseases relate to the digestive system and feed ingredients. These diseases include bloat, diarrhoea, deficiencies or excesses of various nutrients and intake of poisonous chemicals and plants. Examples include:

Deficiencies

Deficiencies of iron, copper, B vitamins and proteins, can all cause anaemia. On the other hand, too much copper can be toxic to the animal as can intake of strychnine or PCB (polychlorinated biphenol).

Poisonous Plants

Animals generally avoid toxic plants if other feed is available. Some plant poisons are:

Goitrogens in the Brassica group (cabbage, rape or canola) suppress thyroxine and cause the thyroid gland to enlarge.

Gossypol in cottonseed; its ill effects on the heart can be reduced by feeding ferrous sulphate.

Lectins (haemagglutinins) and **trypsin inhibitors** reduce protein digestibility in soybeans and other pulses. Cooking before feeding reduces the effect of these compounds.

Mimosine in the tropical forage plant *Leucaena leucocephala* restricts the use of this high protein forage to no more than 10% of the diet dry matter for non-ruminants.

Oxalates precipitate calcium in the blood which then does not allow the blood to clot. Removing the calcium will also induce muscle tetany and paralysis. High oxalates are found in rhubarb, amaranths, wild pigweed (fat hen) and some other plants.

Saponins are found in leguminous forages, imparting a bitter taste. Bloating in ruminants from eating fresh legumes is due in part to saponins. They also reduce the feed intake and growth of baby chicks.

Cyanogens such as hydrocyanic (prussic) acid (HCN) interfere with the oxygen-carrying ability of the blood and can result in rapid death. HCN develops in the plant when, for some reason, photosynthesis of the carbohydrate that is necessary for protein formation does not keep pace with the nitrogen uptake from the soil. Sorghum plants including Johnson grass seem to be the most susceptible to this action.

Alkaloids produce death from severe digestive disturbances, from pain and nervous symptoms. Animals usually die in convulsions.

Some poisonous plants from which animals should be protected include azaleas, buttercups, wild cherries, castor beans, hemlock, some members of the laurel and lily families, milkweed, mistletoe, some mushrooms, oleanders, philodendrons, and potato and tomato plants under some conditions. Although different species of animals often have different tolerance levels for these toxins, care should be exercised in all cases where animals are exposed to them.

Poisons That Might Develop in the Feed

Dicoumarol is produced when the coumarin in legumes is fermented by mould. Dicoumarol is an anti-vitamin K and prevents blood clot formation resulting in internal bleeding. Warfarin rat poison works on this principle. Surgical procedures should be avoided when animals have been grazing sweet clover and several other green pasture plants due to an increased susceptibility to haemorrhaging. When surgery is anticipated, as a precaution, animals should be fed dry grass or hay for the previous 24 hours.

Mycotoxins are produced by mould growth. Aflatoxin is a powerful carcinogen produced by *Aspergillus flavus* and *A. parasiticus* which grow on high moisture, damaged grain and protein supplements such as cracked corn and cottonseed meal. Control can be achieved by keeping feed dry and undamaged.

Ergot is produced by *Claviceps purpurea* which infects seed heads of many grasses such as rye, oats, wheat and Kentucky bluegrass. Ergot causes vasoconstriction leading to gangrene and abortion.

Toxins

Toxins resulting from the fermentation of feedstuffs in the rumen or caecum by *Clostridia* organisms cause enterotoxaemia or 'over-eating' disease. Any changes to high grain diets should be made gradually, and sufficient roughage intake maintained.

GENETIC DISEASES

Malformations and malfunctions can be due to hereditary defects. Crooked toes in chickens, malocclusions of front teeth in rabbits and hermaphroditism in goats are examples.

MECHANICAL INJURY

The injuries resulting from mechanical damage such as broken bones and cuts can cause livestock losses. Stress factors can induce animals to harmful behaviour such as tail biting in pigs, fighting and hair-pulling in rabbits, and cannibalism in chickens.

DISEASE PREVENTION

Disease prevention is easier and cheaper than disease treatment and cure, and sanitation is the keystone to disease prevention. Most pathogens enter the body through a body opening such as the mouth, nose, eyes, genitals or through a wound, although some, including internal parasites, can penetrate the intact skin.

The following preventive and sanitary measures are recommended:

1. Keep animals in thrifty condition so that their natural resistance will be maintained.
2. Provide adequate space free from protruding nails, loose boards or other objects that could cause injury.
3. Group animals according to ages. By keeping younger separated from older animals they will not be contaminated with diseases carried by the older ones. There are other advantages to this management practice as well.
4. Introduce only healthy animals onto the farm. For 3 weeks isolate (quarantine) new animals or those being returned to the premises before allowing them to mix with other animals. Clean and disinfect premises between groups of animals.
5. Sick animals should be isolated from other animals and in some cases destroyed and buried or burned.
6. Vaccinate animals against diseases to which the animals are likely to be exposed and for which there are vaccines available.
7. Internal parasites can be controlled by proper management and by concurrent treatment of all animals with broad spectrum anthelmintics such as coumaphos, phenothiazine and thiabenzole. Ivermectin is an active material produced by a species of fungus, *Streptomyces avermitilis*, found to be effective against both internal and external parasites such as lice and grubs. Similar control can be achieved by treating with both levamisole for internal parasites and famphur pour-on for external parasites. Pesticides can be administered in injections or drenches, and/or in feed, water or mineral blocks. As with most drugs, continuous application reduces their effectiveness, usually because the pathogen or parasite develops a resistance or immunity to them.
8. Institute a good sanitation programme to include the following:
   a. Protect animals from the elements; avoid environmental stress.
   b. Regularly dispose of manure to prevent flies breeding and reduce parasite infestation.
   c. Separate small animals such as rabbits and chickens from their manure and from the soil by maintaining them in cages. Prevent contact with the intermediary hosts of parasites.
   d. Rotate pastures to prevent the buildup of larvae population and to improve grazing.
   e. Control brush and weeds; drain or fence off moist, wet pasture and pen areas.
   f. Provide suitable feed and water containers to prevent contamina-

tion by manure and urine. Empty and clean containers that are so contaminated.

g. Do not offer spoiled feed; remove each day the uneaten, wet, green or succulent feedstuffs to prevent decomposition and mould development.

h. Control rodents, wild birds, insects and other pests.

i. Burn or bury carcasses of animals that die.

j. Practise cleanliness with the liberal use of soap and water.

k. Make effective use of disinfectants.

l. Ultra-violet rays of the sun destroy infectious agents; however, u-v light does not penetrate the surfaces of soil, hessian, etc. very deeply or pass through ordinary glass.

9. Provide adequate ventilation without draughts, and avoid stagnant wetness.

10. Be observant, noting any unusual behaviour that might indicate sickness such as missed feeding or drinking, listlessness, abnormal faecal appearance, rough haircoat, or unthriftiness.

DISEASE MANAGEMENT

The treatment of a disease varies with its cause and symptoms. To combat specific pathogens there are often antibiotics and chemicals which might be taken internally or applied externally as prescribed by a veterinary surgeon. Diseased animals should be made as comfortable as possible and supplied with palatable feed and adequate water to allow the body to heal itself.

Administration of Medication

1. If medicine is given to the animal in its feed, the medicine might be purchased as a premixed commercial medicated feed; or it can be prepared at home by mixing the medicine thoroughly in a small portion of feed—about one-sixth of the daily ration. This is then offered to the animals after a 6 to 8 hour fast. Assume some loss, so overdose by approximately 10%.

2. To administer the medicine in the water, the medicine must be water soluble and the water intake of the animal undisturbed. Only feedstuffs with low water content should be fed while administering medications in this way. As in all cases, carefully follow the directions of the manufacturer.

3. Direct administration of medicine allows exact dosing and dispensing of medication, especially when feed and water intake is erratic or disturbed, and where only individual animals are to be treated.

A goat can be induced to swallow a bolus (large tablet) or capsule or even a dry powder that has been wrapped in paper. Back the animal into a corner and while pressing against it with your body enclose the upper neck with one arm and insert the thumb in front of the premolars. Prise the mouth open while lifting the head. Place the bolus in the mouth on the rear of the tongue either with the hand or with a 'balling gun'. Substances in this area will stimulate a swallowing reflex. Exercise care to keep the fingers from between the molars and premolars. In larger animals it might be more convenient to grasp the tongue and restrain it outside the mouth until the medicine has been administered.

Liquids, perhaps up to 500 ml, can be administered to larger animals with a dose syringe or with a narrow necked container such as a soft drink bottle. The animal's mouth is opened as described above and the neck of the bottle inserted about 5 cm into the corner of its mouth. The head is elevated and the contents poured into the back of the mouth.

A stomach tube can be used for medication with larger quantities of liquids. Choose a tube of the size appropriate to the animal and curve the end of the tube upward until through the pharynx into the oesophagus. Be aware of the location of the stomach in order to know how far to insert the tube. Care must be taken not to pass the tube into the trachea because administration of liquids in the lungs produces pneumonia. After the tube has been inserted, test for proper positioning by listening for breathing sounds, and by pouring a small amount of clean water into the tube. If the water enters the trachea, the animal will cough intensively and show great uneasiness and distress. When properly placed, using a funnel or similar device facilitates pouring the medicine down the tube. Exercise care that the animal does not chew off the tube and swallow it.

Medicines can be injected using a hypodermic needle. Syringes and needles must be clean and preferably sterilised before use. Injections may be made beneath the skin (subcutaneous), within the muscle (intramuscular), within the abdominal cavity (intraperitoneal), or if a quick response is needed, injection is made into the bloodstream (intravenous). Intravenous injection is usually in the jugular vein in the neck of a goat and in the marginal ear vein of the rabbit.

Vaccination is applied before the disease strikes in order to stimulate the animal to develop immunity to the disease. There is no merit in vaccinating an animal to develop immunity for a disease which it has already contracted. However, the course of some diseases can be lessened by the injection of a specific antiserum which contains antibodies produced by another animal. Antiserum, antibiotics and various drugs can be injected into the animal body to combat a disease already contracted.

Drug Dose and Tolerance

The manufacturer's directions should be followed in determining how much of and the manner in which the medicine should be administered. These amounts are generally based on the body weight of the animal.

Signs of intolerance or reaction to the drug dictate immediate interruption of treatment. These signs might include profuse salivation, cyanosis, respiratory distress, diarrhoea, daze, disordered equilibrium, hair loss, subcutaneous bleeding and abortion.

To learn if a new drug will be tolerated, first test the effect on one animal only. (This is not completely reliable because one animal might tolerate the substance and another might not.) Those drugs most likely not to be tolerated are unfamiliar antibiotics and preparations containing glycosides and alkaloids.

Tolerance to certain medications might be lower in pregnant or lactating animals.

Drug Residue

Consider the likelihood that active substances are transmitted to the foetus and milk and eggs. Evaluate their suitability as food.

Meat animals must be free from drug residues at slaughter. It usually needs at least 3 days for drugs to clear the animal tissues. For withdrawal time always follow the instructions accompanying the drug package.

Post-Mortem Examination (necropsy)

Every animal that dies without a known cause should be necropsied. Ordinarily such operations are performed by veterinary surgeons who are trained not only to recognise diseases but to protect themselves from possible infection. In the absence of such qualified persons, over a period of time, performing necropsies will educate the herdsman as to the appearance of both normal and diseased organs and tissues.

The necropsy should be performed in an area which provides enough light so that small spots or other abnormalities can be detected. Body fluids and tissue debris should be contained so they can later be disposed of. The area should be one that can easily be cleaned and disinfected after the examination.

The person performing a necropsy should have a sharp knife (or scalpel), a good pair of scissors, bone-cutters (or saw) for large animals, a pair of rubber gloves, and a few specimen bottles with 10% formalin or other preservative. These bottles can receive samples of diseased tissues that might be sent to a diagnostic laboratory for evaluation, or kept for later comparison. Entry in a permanent record book should be made of all examinations, even if the findings were negative.

Before cutting into the dead animal, carefully examine it for any external abnormalities or parasites. Then place the animal on its back with head and legs extended. Make an incision along the midline from the chin to the hind legs cutting through the skin and underlying muscular wall. Exercise care not to puncture the internal organs. Cross incisions can be made at the hind legs and behind the rib cage to better expose the abdominal viscera. Note if there is excessive fluid and systematically examine the abdominal contents. Remove the liver, examine its surface and make cuts in it to examine the interior; this includes the gall bladder. Do the same with the spleen and each kidney. Then look closely at the stomach, small intestine, caecum and large colon. Note any abnormalities in condition, colour or size. Then remove and cut open and examine the inside contents and mucosal lining of the gastro-intestinal tract. Note any inflammation, impaction or parasites.

Next, open the chest cavity by splitting the sternum and pushing back the ribs. This will expose the lungs and heart. Any extra fluid, inflammation or other abnormalities should be observed. The windpipe from the pharynx to the lungs and the bronchi should be slit open and inspected for anything not typical. The interior of the heart and lungs should be similarly examined.

After completing the necropsy, the debris and carcass should be burned or buried deeply enough that it would not be dug up and scattered about by scavengers. Covering the carcass with lime is a good practice. The area and all equipment should be cleaned and disinfected.

DISINFECTANTS

The maintenance of animals in large numbers or on the same ground through successive generations results in a buildup of disease-producing viruses, bacteria, fungi and parasite eggs. The disease cycle can generally be broken by keeping susceptible animals from the area for extended periods of time and by tilling the land. However, periodic, thorough cleaning and disinfecting of all buildings and equipment might sometimes be more practical.

Because organic matter, as in dirt and manure, inactivates most disinfectants and protects microorganisms, a good cleaning of buildings and equipment should precede disinfection. The use of a stiff brush with hot water and a cleansing agent followed by rinsing with clear water are recommended. Allow the buildings and equipment to dry.

Factors to consider in choosing a disinfectant include:

1. Effectiveness in the presence of organic matter.
2. Compatibility with soaps, or detergents, or water hardness.
3. Harmlessness to building materials and equipment.
4. Relatively non-corrosive and non-toxic to man and animals.
5. Effectiveness under conditions of temperature range and acid-base reaction that will be encountered.
6. Spectrum of activity (whether it destroys gram-positive or gram-negative bacteria, tuberculosis bacilli, bacterial spores, fungi and/or viruses).

Types of disinfectants:

1. Phenols (carbolic acid, cresol) being essentially insoluble in water are frequently combined with a soap (saponated, e.g. Lysol) to increase solubility. Cresol is applied in a 2% to 5% solution; 35 g cresol per litre of water. If the solution is hot it is more effective. Saponated cresol is one of the best disinfectants to use in the presence of organic matter, and it is non-corrosive. However, its strong, persistent phenolic odour makes it objectionable in tightly enclosed buildings and with equipment that will come in contact with food. Cresol is effective against most pathogens but not against spores. Some coal tar derivative compounds are thought to be carcinogenic; therefore, one should avoid prolonged body contact with the material.

71

Synthetic phenols (e.g. orthophenylphenol) retain advantages without the disadvantages of cresol.

2. Iodine crystals (2% or 7%) dissolved in alcohol make a tincture and can disinfect surface areas such as cuts and scratches, ringworm and be used for dipping the navel of newborn animals.

3. Iodophores are combinations of iodine and agents that aid solubility ('tamed iodine'). They are non-staining, non-irritating and largely free from inducing skin hypersensitivity. Their effectiveness is reduced by alkali soaps, but they can be used in disinfecting clean utensils and equipment and living tissues.

4. Chlorine compounds, as in laundry bleaches, being strong oxidising agents, have rapid action against bacteria, spores, fungi and viruses on cleaned surfaces. A 2% solution of calcium hypochlorite (bleaching powder, chloride of lime) is a cheap but effective disinfectant and deodoriser.

   Chlorine is corrosive. Since the chlorine dissipates into the air, these compounds should be stored in air-tight containers.

   Chloramines are organic chlorine compounds which release chlorine slowly and exert a prolonged bactericidal effect. They are less toxic and irritating than the hypochlorites.

5. Lye (soda lye, caustic soda, sodium hydroxide) is highly disinfectant as a 5% solution (one 380 g can in 75 litres of water). A less hazardous and irritating solution is made with 20 g lye in a litre of water. Lye, a caustic poison, will damage painted or varnished surfaces, textiles and aluminium. It is not effective against tuberculosis and spore-forming bacteria.

6. Quaternary ammonium compounds are surfactants commonly used for general disinfection of dairy, meat packing, food handling, and eating and drinking equipment that has been pre-cleaned and rinsed. They are antibacterial but largely ineffective against viruses, fungi and spores, and are inactivated by organic matter.

7. Formaldehyde and other aldehydes are effective disinfectants, especially as fumigants. Formalin is a 40% solution of formaldehyde. A 4% formalin solution (or 1% to 2% solution of formaldehyde) is effective in 15 minutes against viruses, tuberculosis organisms and anthrax spores. This solution is also effective as a footbath in the control of hoof rot. Both formaldehyde and paraformaldehyde are used as fumigants for about 8 hours in tightly closed buildings. Glutaraldehyde is a more effective germicide with a less irritating odour than formaldehyde, but it is more expensive. In any case, care should be taken in the use of these fumigants because of the damage they can do to both humans and animals. Formaldehyde is a popular ingredient of embalming fluid.

8. Chlorhexidine is an excellent commercial product used to disinfect surgical instruments and operation sites. It is effective against both viruses and bacteria.

Caution   Many cleaners and most disinfectants are poisonous. They should be stored in tightly closed containers in an area not available to children and unauthorised personnel. Keep labels on the containers and carefully follow the directions for use. Observe safety precautions, avoid skin contact and inhalation of gases and wear protective goggles and gloves when handling concentrates and hazardous materials.

GENETIC CONTROL OF DISEASE

There are a limited number of diseases to which certain animal species or breeds have developed more or less resistance through natural selection. However, selecting animals to increase genetic resistance to infectious disease is usually thought to be non-productive. However, it is wise to avoid indiscriminate inbreeding, and breeding animals with known anatomical weaknesses, abnormalities and nervous disorders.

PEST CONTROL

Wild birds, rats, mice and other pests not only spread many infectious diseases but also do other damage such as property and feed destruction and contaminating feed and equipment with urine and faecal droppings.

Birds can effectively be excluded by covering windows and other openings with wire netting. Sparrow, starling and other nests should be removed as they are built. Bird numbers may be decreased by shooting, trapping or poisoning.

Rodents, especially rats, create special problems. Rats are estimated to eat or contaminate 42.5 million tons of grain each year, enough to feed 200 million people. In the USA the expense of each rat is thought to be near $30 annually. They spread many human and animal diseases including typhus, tularaemia, trichinosis and pseudorabies as well as external and internal parasites. They kill baby chicks, start fires by gnawing through wire insulation and weaken building foundations by burrowing under them. One pair of rats produce 75 to 100 offspring each year in 6 to 10 litters of 8 to 12 offspring which mature and start reproducing on their own in 3 to 4 months.
    Rodent control includes:

1. Anticoagulants such as warfarin are mixed with grain at 0.025% (0.05% for quicker results with mice). By chemically tying up the vitamin K, warfarin prevents the liver from producing prothrombin which is necessary for blood clotting. The poison is then responsible for death through haemorrhaging. The anticoagulant is tasteless and odourless and cannot be detected by the animal. Multiple doses over 5 to 14 days are required to induce fatal haemorrhaging. The poisoned rats die in their burrows or nests. Rodents do not associate the death with the bait.
2. Shallow containers with about 1 cup of mixed bait are placed in darkened areas frequented by rats. Keep bait dry and check daily to replenish the supply.

3. Mice should have more bait stations with less bait. Mice control may take longer because they eat less and are more resistant than rats.
4. Keep storage areas and premises clean and orderly; burn trash and rubbish piles; store wood, lumber, boxes, sacks, etc. 18 to 24 inches off the ground. Grain in bags should be stored in rodent proof bins, if possible; close all foundation openings with metal shields.
5. Keep children and pets away from the bait. Even though the level of toxicity is low, warfarin and other anticoagulants can be fatal if enough is consumed over a long enough period.

6
ANIMAL SELECTION AND BREEDING FOR GENETIC IMPROVEMENT

SELECTION

Animals serve many significant purposes in agriculture. Before introducing an animal into a farm situation, it is important that the role the animal is expected to play is decided upon. Then the species, type and individual animal can be selected that will best fulfil these expectations. The capability of meeting the feed requirements and providing the animal freedom from environmental stresses must be evaluated. Animals must be able not only to survive but also to produce under the conditions of climate, nutrition, sanitation and management to which they are subjected.

POWER

If animals are to produce power they must be of a temperament with sufficient size, strength and stamina to respond to man's wishes. While the ass will generally meet these criteria, cattle and buffalo have the added advantage of producing milk and providing a meat that is more widely acceptable for consumption. Power provided by such a large animal could be of real benefit in increasing production but could be introduced only if there were sufficient production potential to produce sufficient feed in excess of family needs to meet those of the animal.

Bullocks of both buffalo (*Bubalus bubalis,* not to be confused with the North American bison that cannot be trained) and cattle are capable of draught. Compared to cattle, buffalo are often more docile and produce milk with a higher fat content, but milk yield potential is more variable. The buffalo have a lower breeding efficiency, poorer viability of young and are slower maturing. In the tropics cattle are generally 3 to 4 years old before dropping their first calf; buffalo are older still.

FOOD

In providing additional nutrients, managing a few, smaller sized animals is less difficult than trying to raise large animals. The meat from small animals does not require refrigeration or extensive preservation because the entire animal can be consumed by the family in a few meals. However, goats, rabbits and guinea pigs can be utilised only where forages are available.

Chickens

Chickens have found acceptance in essentially every country of the world. There is widespread availability of good breeding stock. The farmer must differentiate between laying and meat types and select the type desired. If adequately fed and protected from predators and diseases, they can survive, reproduce and produce both nutritious eggs and meat. Data in Table 3 show the laying hen as being the most productive in producing energy and protein. Chickens can serve as scavengers and survive and produce on a wide range of feedstuffs. If a feed for animals cannot be raised but must be purchased, chickens have a distinct advantage over other animals because of their efficiency of converting feed into animal products and the more widespread availability of chicken feed in the market place.

Table 3   Yields of Different Types of Chickens

Factor	Layer	Dual Purpose	Meat Broiler	Meat Roaster	Meat Roaster
Time span (weeks)					
Growing	20–22	20–24	8	12	16
Laying	48–60	38–50			
Optimal conditions					
Eggs produced	240	160	—	—	
Liveweight (kg)	1.81	2.72	1.81	2.72	3.62
Reasonable production expectation, annually*					
Eggs (number)	160	107	—	—	—
Meat (kg)	0.91	1.18	5.22	5.22	5.22
Food produced*					
Energy (kcal)	14,940	11,235	9100	9100	9100
Protein (g)	1140	835	690	690	690

* Based on the assumptions:
1. The nutrient content of meat and eggs is:

	Energy (kcal)	Protein (g)
Per kg of fryer meat	842	126
Per kg of broiler meat	1744	133
Per kg of roaster meat	2176	127
Per kg of eggs	1630	129
Per egg (50 g)	81	6.4

2. The operation is continuous at two-thirds optimal conditions, immediate replacement of chicks and a one-third slaughter loss in blood, feathers and viscera. Layers are 18 months old when slaughtered.

Rabbits

Since rabbits are herbivorous they need not be in competition with man for food. In fact the greatest potential for using the rabbit for food lies in areas where forages are readily available, grain is expensive and limited and where the need for human food is great. Their being able to eat kitchen and garden wastes, leaves and weeds makes rabbit meat and wool production possible on very limited farm land.

Rabbits, in providing meat and pelts, fit in well with the small farm programme. About 80% of the carcass is edible, and is of the same white meat type as poultry and veal. Being small, the entire carcass can be consumed in one meal by a family; this eliminates any need for preservation by refrigeration, drying or curing. Rabbits have demonstrated an ability to reproduce and survive under local management conditions in tropical as well as temperate environments. Under good management, doe rabbits can begin to reproduce at 5 months of age, require 31 days for gestation, and can be rebred 2 to 5 weeks after kindling. Each doe should produce 4 litters per year, a total of over 20 offspring, yielding 60 kg of live meat (32 to 50 kg of dressed meat). It has been estimated that 1 buck and 4 does each producing 4 litters of 8 kits can, in those 16 litters, produce more meat annually than an average beef cow, and this on far less total feed. Stated another way, the offspring of a female rabbit might well weigh more than 10 times her own body weight.

Rabbits can convert 3 kg of a balanced ration into 1 kg of growth as compared to a conversion efficiency of 2.5 in table poultry and 8.1 in steers. (To get a true picture when comparing these efficiency statistics, the nature of the diet should be considered.) Expressed another way, the energy expended by rabbits per kg of weight gain is 39% below that of sheep and 47% below that of cattle while producing 2000 kg of liveweight per hectare. Where the concentrated feed for chickens is relatively expensive and difficult to get, rabbits eating locally produced plant material are more economical. Limited needs for cash make rabbit farming accessible to even the poorest segments of the rural population.

Raising rabbits is a labour intensive operation but can make profitable use of family labour. The light nature of the work is such that it can be done by women and children. Rabbit farming can be carried out on various scales, depending on land, labour, capital and managerial resources of the farmer. Undertaken on a small scale with average management, rabbit farming can provide substantial additions of nutritious meat to the family diet. Under better management conditions and undertaken on a larger scale, it could provide significant cash income.

Due to their relatively small size, the rabbits' feeding and housing requirements are small. Their successful production is dependent upon an adequate supply of roughage; year round forage production would be helpful.

Constraints on rabbit farming include:

1. Sufficient experience and commitment to regular animal care are required for successful production. Rabbit production requires a high level of

management skills, rabbits being generally harder to raise than chickens. Rabbits are almost totally dependent upon man for feed and care. Furthermore, they are relatively highly susceptible (especially when young) to diseases of the digestive and respiratory tracts.

2. Quality breeding stock may be difficult to locate.
3. Rabbit farming may not be a traditional practice, therefore extensive promotion, training and monitoring will be needed for its successful introduction in some areas. This training should be directed toward those who will be caring for the animals, usually the women and children.
4. Animals must be reared in confinement, protected from inclement weather and direct sunlight. High temperatures (above 30°C) reduce reproduction and growth. Because of its lack of a loud reaction to threatening situations the rabbit is especially endangered by human and animal predators. This requires at least daily inspection and care, a condition that for some would be difficult to meet.
5. Land and/or other resources must be allocated for growing and storing feed. It has been estimated that harvesting forage can require 30 minutes per kilogram of meat produced.
6. Intensive labour requirements include bringing feed and water to the animals. Each breeding animal requires individual caging, care and attention.
7. Sociological/cultural attitudes may hinder rabbit raising. There is a reluctance on the part of some to eat rabbit meat. Some people refuse to do so because they consider that rabbits are pets and should not be subject to slaughter and consumption. Others erroneously think of rabbits as rodents when actually they are an independent order of hares (*Lagomorpha*). Orthodox Jews discriminate against all rabbits, others refrain from eating white rabbits because they are considered holy.
8. Lack of reliable markets may present problems for commercial production.

Goats

If culturally acceptable, goats make desirable additions to the small farm economy. Goats are nimble, hardy and adaptable, well suited for uncommercial, limited resource farms to supply a small but rather constant supply of milk and meat. There are several goat breeds that have been bred to have highly developed mammary systems together with a capacity to eat sufficient feed to produce large quantities of milk; many individual does have produced over 1000 kg of milk in one year. Such production depends not only on the animal's genetic capacity but also on an adequate and balanced diet, freedom from disease and environmental stress and on skilled management. The lack of any of these factors reduces the ability of any superior heredity to express itself, resulting in lower production levels.

In the tropics the value of a goat for milk has been generally considered secondary to meat production. Compared to the cow, an individual goat requires less labour and feed and can be more easily cared for by women and children. Goats are browsers, preferring leaves of forbs and trees to the grazing of grass. Their higher threshold for bitter taste gives them an ecological niche in which they can help to control brush encroachment and eat weeds and certain other of a wide range of plants avoided by other animals. Being browsers, goats are thought to be less likely to pick up the heavy worm infestation burdens that the grazers of short grasses do.

Scientific evidence is lacking to show that goats have fewer health problems than other species. Although goats have developed in a hot desert environment, they can adapt themselves to a wide range of climatic conditions. Goats are found in both cold and hot areas but generally do not adjust as well to moist atmospheres. In searching for feed, non-dairy goats can travel long distances requiring infrequent watering. While adequate fresh water should be provided at all times for all animals, in winter in temperate areas when sweating and respiratory cooling are unnecessary, some non-lactating goats can obtain their water needs from their feed (especially if feed has more than 60% moisture) and from metabolism. At high temperatures their water turnover rate is greater than for the camel but 11% lower than sheep. In many indigenous goats, protection from the radiant heat of the sun is provided by reflecting coats of short, fine, glossy, medullated hairs which lie flat against the skin. Indigenous tropical breeds have been selected:

1. to tolerate solar radiation and other climatic stresses,
2. for enhanced resistance to local diseases and infections, and,
3. so as to require less feed, and thus they are genetically geared to low production.

Capacity for dry matter intake is a determinant of productivity. Intake is influenced not only by the temperature and level of production, but also by the nature of the feed (succulence, palatability, ratio of roughage to concentrate, etc.) and health. Daily dry matter intake is about 3% to 5% of the adult body weight in goats.

Being a seasonal breeder is a serious handicap, yet goats do have a high rate of reproduction. They can kid first at 1 year of age after a 5 month pregnancy. The twinning rate is near 40%, providing adequate feed is available. Since the shortening of daylight (photoperiod) induces the breeding season, goats in areas closer to the equator have a less defined breeding season, tending to mate throughout the year.

Constraints on goat production are related to their grazing habits and size. If grazing is not controlled serious ecological damage can result. Goats cannot pull a plough and their labour and management costs are high in relation to their individual value. Also, they are susceptible to predation and theft. Their agility and curiosity make confinement of goats more difficult. Changing from a meat to a milk producing type requires a concurrent step-up in feed and management skills.

Other small ruminants include sheep, and the genus *Lama* (llama, alpaca, vicuna and guanaco).

Guinea Pigs (Cavies, Cuis)

Guinea pigs are popular small animals in some areas of the world. They readily reproduce, are tractable, easy to feed, house and handle indoors, and can even be reared in cardboard boxes in city apartments. However, they have a poor feed conversion ratio and the number of offspring per gestation is small. Improved breeding stock is usually difficult to obtain or not available.

Additional Animals

Other food-producing small animals that might be considered are: pigs, capybara (the largest rodent in existence found in South America and characterised by partly webbed feet, no tail, and coarse hair), iguana (a herbivorous reptile characterised by a dorsal crest of soft spines and a dewlap), pigeons, quail, ducks, geese and fish.

OTHER USES

Small animals may be maintained to produce hides and pelts, hair or wool, fertiliser or for companionship or other aesthetic reasons.

BREEDING GOALS

The aim of a breeding programme should be to develop utility in the environment in which the animals must live; animals with high genetic potential for production and thus higher management requirements should not be introduced into regions where their owners cannot hope to meet their nutritional, health and other environmental requirements. Improved breeds of animals imported to unfavourable environments may prove inferior to indigenous (criollo) breeds.

Lifetime performance is the only criterion by which the economic utility and constitutional vigour of animals can and should be judged. Selection factors for improvement of breeding animals should be:

1. Survival ability.
2. Economic productive capabilities in various and often hard environments.
3. The ability to transmit any desirable characteristics to their offspring. It is not the success of the imported animals that counts, but rather the lifetime performance of their progeny.

Limits to genetic improvement include the following factors:

1. Livestock may play only a minor role in the overall economy; therefore, priority of concern and time is given to other things.
2. There is insufficient land to produce adequate feed for animals.
3. Lack of feed supplements to add to locally produced feeds to obtain a

balanced ration to allow achievement and observation of genetic potential. Malnutrition and other aspects of a poor environment mask the genetic ability to produce. Many productive traits are influenced more by environment than heredity. They have low heritability estimates.
4. Availability of capital to buy improved stock or the equipment and facilities for animal management are frequently lacking.
5. Inadequate knowledge of the principles of animal breeding, feeding and management limit successful production.
6. Other aspects of a poor environment and management restrict the expression of superior breeding.

It should be recognised that small farmers expect their animals to serve several purposes. They may be expected concurrently to work and to produce milk, young animals, organic fertiliser and meat. These objectives may not only conflict with each other but, by selecting for several characteristics at the same time, they make improvement in any one trait occur slowly, at best.

Before exotic improved breeds can be successfully introduced, environmental limitations must be overcome, such as: disease, inadequate nutrition, exposure to a harsh environment (e.g. heat) and failure to fully recognise animal needs (e.g. a high producing goat will need to be regularly milked twice daily as well as being provided with adequate feed and water and protected from a harsh environment).

If these limiting factors cannot be met and controlled, the possibilities for genetic improvement within the native stock should be exhausted before trying to introduce exotic breeds.

BASES FOR INDIVIDUAL SELECTION

There are 3 bases used in selecting individual animals:

1. Body type or conformation; what the animal looks like. Unfortunately, this is usually the only data available for selection. A desirable animal is one of appropriate size for its breed, giving the appearance that it is capable of doing what it is being kept for, whether it be meat, milk, eggs, pelts or wool production. It should be sound, free from apparent diseases and defects such as wryness, poor feet, worn teeth, hernias, cryptorchidism, nervousness, etc.

The capacity to consume feed is important. Productive performance depends on adequate feed intake. Forage intake depends on the size of the abdomen and body weight. In milk producing breeds, larger animals produce more milk if their size is due to scale and not to fat or muscle mass; this disfavours the selection of these animals for meat production. The size and pliability of the udder reflects milk secretion capacity. Well-attached udders are desirable.

It is unknown what body types (if any) have a bearing on adaption to adverse environmental conditions and utilisation of coarse feeds. Healthy, well-formed feet and legs are important, especially for grazing animals.

2. Pedigree provides not only names of ancestors but, if complete, provides type and production information on the animal and on the parents, sibs, half sibs and other relatives.
3. Performance is the most useful tool in deciding whether to add or retain an animal in the herd. Productive performance includes growth rate, milk production, litter size, etc. The performance of offspring in comparison to their dams provides a useful technique for evaluating the genetic value of sires. In rabbits the survival of rabbits born is correlated with the ability of the doe to make a good nest.

SOURCES OF VARIATION

Every individual animal is unique. This uniqueness is due to differences in its heredity, the environment to which this heredity is subjected and the interaction of heredity with environment. The genetic variation in production traits is related to tissue enzymes, hormones and various intracellular processes. Variation is an important component among animals; without it there could be no selection for improvement.

HEREDITY

Nature of genetic material Heredity is a function of the many genes in an individual. Genes determine the size, species, breed, shape, productive capacity and the many other characteristics that are transmitted from parents to offspring. These determinants are situated in specific locations (loci) on one of a chromosome pair. The number of these chromosome pairs is characteristic of the species; and members of these pairs are said to be homologous. One of these gene-bearing chromosome pairs comes from the sire via sperm and the other from the ovum produced by the dam. Thus, half the genetic material comes from the sire and half from the dam. Genes occupying the same locus on a pair of homologous chromosomes are called alleles. Whereas the alleles affect the same characteristic (e.g. hair colour) they may affect it in different ways. If both alleles affect a character in the same way they are said to be homozygous; if not, heterozygous. If one heterozygous allele masks the effect of another, it is said to be dominant over the recessive gene.

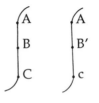

The above chromosome pair shows genes A, B, and C and their alleles A, B′, and c. This individual would be homozygous for the dominant gene A but heterozygous for genes B and C; c is recessive to C.

Deoxyribonucleic acid (DNA) is the basic material of heredity. It has the unique property of being able to reproduce itself. It is the material of which genes are made, thus it directs the life processes of development and performance.

Genetic Improvement

There are only a few characteristics important in animal production that are influenced by only one or two genes (horned or polled, red or black, etc.). The characteristics of economic importance (growth rate, body size, reproduction rate, disease resistance, milk or egg production, etc.) are controlled by a multiplicity of genes. The object of the animal breeder is to increase the frequency with which the desired genes occur in the population, breed or herd.

The geneticist comes to think in terms of averages (means) of populations rather than the appearance or performance of any one individual animal.

A time-proven adage in livestock improvement is to breed the best to the best. It has similarly been observed that the average of the offspring of improved parents is usually nearer to the average of the population as a whole than to the average of the parents. This is a phenomenon referred to as **regression**. The extent to which the offspring average varies from the average of the population toward the selected parents is called heritability. **Heritability** is a measure of the relative importance of genetics and environment in deciding the nature of individuals. Consider a particular trait like rate of gain; if 30% of the variation among individuals is due to heredity and 70% due to environmental influences, then the heritability is 30%. For example, if the parents gained 100 g daily and the population average was 60 g, the offspring average would reflect an improved 30% of 40 g or 12 g daily gain. Assuming uniform environmental conditions the average gain of the offspring would be 60 plus 12 or 72 g daily.

High heritability of a trait means that rapid genetic improvement can be made by selecting for that trait. Or, stated in genetic terms, heritability is the phenotypic differences in a trait that can be attributed to inheritance. Some characteristics are more heritable than others, as indicated by the heritability estimates that have been published for the various classes of livestock. For many production traits one might expect a heritability of 25% to 30%, meaning that superior parents will transmit to their offspring only 25% to 30% of their own superiority over that of the breed. The other 70% to 75% of their superiority is due to the environment to which they were subjected and the effect that this environment had on them. If the trait has high economic value one might still apply selective pressure even if the heritability is low. For example, we would select better laying hens and more prolific sows and does to be the parents of the next generation even though less than 10% of this superiority would be transmitted to their offspring.

Effects of Selection

The genetic makeup of the herd is changed by selection. The process of selection is to choose the animals that are most desirable, then manage the flock so that these better animals will produce more offspring for the next generation. Genetic capability can be masked by a poor environment; therefore, animals should be managed so that genetics become the limiting factor, not feed, disease or other environmental factors. Furthermore, the greatest genetic improvement can be realised when selection is being made for the fewest possible traits. The more traits for which selection is being made, the lower the selective pressure and the slower the rate of improvement in any one character. For example, it is impossible to save a doe for her high milk production while culling her for her pendulous udder.

In summary, quantitative characteristics such as production of milk or eggs, litter size, growth rate, body size, stamina, etc. all require the complex interaction of a great number of genes. The genes determine not only the rates at which endocrine glands secrete their hormones, but also the sensitivity with which the end (target) organs respond to the hormonal or neural stimuli acting on them. The fundamental equation for genetic selection states that the improvement in genetic makeup of a population depends not only on how superior the selected parents were in the selected trait, but also on how much of this superiority is transmitted to their offspring (heritability). Some traits, such as body growth and size, are moderately heritable (15% to 30%); therefore, they can be easily improved by selection. However, others such as the reproductive traits of litter size, hatchability and egg production are less than 15% heritable, which is considered to be low. These are much more affected by management and other environmental factors, and thus selection is not as effective in bringing about genetic gain. Improvement for these traits can be made much quicker by changing the environment.

The genetic improvement in any trait that can be made in a herd depends not only on the heritability of that characteristic but also on the selective pressure (selection intensity) applied to it and the accuracy with which the genetic makeup of the animal is evaluated. The rate at which this improvement is made will also depend on the generation interval or the time elapsing from birth to reproducing.

Progeny records are the most effective means of determining the breeding value of an animal or of evaluating the genetic makeup. The greater the number of records, the more reliable the information. Selection of breeding animals, then, is most effective when it is postponed until the animal proves that it has the desired characteristics. If the desired traits can be measured only after the animal is dead, selection should be made in part upon the performance of sibs and half sibs and parents as well as offspring.

Mating Systems

Once the type and individual animal have been selected to become parents of the next generation, various mating strategies can be used, depending on the desired goals. Both the sire and the dam share equally in supplying the genes that determine the characteristics of the offspring.

Mating systems which might be employed in a herd or flock to improve the genetic base include:

1. Inbreeding is the mating of relatives, such as sire to daughter or brother to sister; crossing like with like. This leads to an increased homozygosity (having more of the same genes). If there is no concurrent selection being practised, inbreeding creates greater genetic extremes in a population. Also, the more homozygous the parents, the truer to type they will breed, but with a resulting loss in size and vitality.

   Inbreeding with its increase in homozygosity permits the 'fixing' of genes in a population; this means that since the animals carry only the desirable genes they will 'breed true'. Vigorous and consistent selection among large numbers of animals is necessary to make significant progress toward a specific goal.

   Inbreeding causes both good and bad, as well as dominant and recessive genes to become homozygous. In the heterozygous condition undesirable (sometimes lethal) recessive genes are not apparent. However, if 2 animals carrying the same undesirable recessive gene are mated, on the average one-fourth of the offspring will not carry the recessive gene, one half will be heterozygous like the parents and one-fourth will display the undesirable characteristic because it is homozygous for the detrimental gene. Thus, inbreeding is useful in concentrating desirable genes from superior lines and also in identifying the carriers of undesirable recessive genes. However, outcrossing should be the mating plan of the small breeder due to the accompanying increased size and hybrid vigour.

2. Linebreeding is mating in such a way as to increase the relationship of the offspring to a desired ancestor; a mild form of inbreeding.

3. Outcrossing is crossing unrelated animals within the same breed.

4. Crossbreeding is mating animals of different breeds (or species). Outcrossing and crossbreeding produce heterosis (hybrid vigour) in which the size, vigour and performance in the offspring is above that otherwise to be expected. However, this increase is greater in the first than succeeding generations. The improved performance is generally due to greater aggressiveness at the feed trough; the animal eats more and consequently produces better. The more diverse the genetic background of the parents, the greater the hybrid vigour of the progeny. Crossing of different inbred strains is a popular application of this principle. Crossbred animals are less likely to transmit their own traits to their offspring; they do not 'breed true'. Generally, traits that have the lowest heritability are those most responsive to heterosis.

The management objective is to provide circumstances so that the animal and its environment are in harmony. Usually, it is quicker and easier to modify the environment than it is to modify the animal. The process of bringing the environment in harmony with the genotype of the animal is called management. Errors in estimating the genotype (heredity) from the phenotype (what the animal looks like) are reduced when the heredity is not masked by an unfavourable environment. With a favourable environment, what the animal can produce is determined by its genetic makeup. Good management includes such things as regularity in feeding, sanitation, breeding on time, prompt treatment of parasites and diseases, and consistent record keeping.

SUMMARY

Any breeding plan to be effective must be accompanied by selection. In fact, selection is the most potent force to change gene frequency, or bring about herd improvement.

The type of animals ultimately selected for use will depend on:

1. Feed supplies.
2. Animal health conditions.
3. Social outlook and economic status of the owners.
4. Marketing opportunities and limitations in the use of animal products which might be affected by different faiths and social customs.
5. Willingness to do the work associated with the more critical care and management of higher producing animals.
6. Availability of breeding stock.

Indigenous tropical breeds usually have a superior tolerance to climatic stress. These breeds have developed an ability to survive rather than to produce. Under some circumstances genes for higher production can be introduced through importing exotic breeds. Intermediate changes in resistance and production might be expected by crossing the endogenous with exotic breeds and then carefully selecting the offspring for the more desired traits.

Part Two
ANIMAL PRODUCTION PRACTICES

GENERAL OUTLINE

Production practices and learning activities follow for poultry, rabbits, goats, guinea pigs and pigs. Concern is given to:

1. Buildings, equipment, space and facilities needed for production.
2. Selecting a species and type of animal and the individual animals.
3. Reproducing the animals—mating, pregnancy, parturition.
4. Nutrition and feeding—the animal nutrient needs and how they can be met.
5. Sanitation and hygiene—disease prevention.
6. Disease identification and control.
7. Controlling the animal and its environment, raising replacements, handling techniques, gathering eggs, milking, trimming hooves, dehorning, castrating, special problems in the care of animals at different periods, adjusting the environment to the animal's needs, etc.
8. Care and methods used in processing the products—milk, cheese, butter and yogurt, and eggs.
9. Animal slaughter, meat cutting and curing, pelt and hide care.

7
POULTRY PRODUCTION PRACTICES

The new post-war poultry industry as seen in the Third World is essentially an urban phenomenon, financed by urban capital, mainly befitting urban-oriented producers and urban consumers and generally located adjacent to the cities and larger towns. It has not greatly touched the subsistence and small-scale poultry producers in the numerous villages throughout the tropics where millions of people still depend upon backyard poultry production for their eggs and poultry meat. The chicken is by far the most widespread, although turkeys, ducks, geese, pigeons and other birds make important contributions to the food supply in various regions.

In the developed countries the chicken has become a specialised machine capable of previously unheard of growth and productivity. This is a consequence of both sophisticated breeding programmes to change the genetic makeup of the birds and of skilful management which provides feed and a disease-free and climate-controlled environment which allows the superior genetics to express themselves. Table poultry have been grown to a 2 kg (4.4 lb) weight in 50 days on 4 kg (8.8 lb) of feed.

Laying hens might begin production at 20 to 24 weeks of age, laying an egg almost every day for about a year. Maximum production is reached at about 6 to 8 weeks after coming into lay; production will then taper off to the end of the laying year when birds normally go into moult.

It should be noted that maximum animal production depends upon superior animals in a suitable environment. It may not be possible to provide such conditions in many small-scale agriculture systems. However, the presence of farm flocks throughout the world demonstrates the possibility of achieving production levels that are acceptable to local growers. Such levels can be increased through better breeding, feeding and management techniques.

Before embarking on a poultry project, family resources, commitment, needs and preferences should be brought to bear in deciding the type of chicken to use, whether primarily for laying eggs or for producing body gain for meat. For the different types of chicken, the approximate yields of eggs and meat in terms of calories and grams of protein are presented in Table 3 on page 76. The data show leghorn type egg laying hens to provide the greatest quantity of energy (14.940 kcal) and protein (1.14 kg) than do either dual purpose or meat types on an annual unit basis.

CHICKENS

Laying Type

Chickens for egg production should have small bodies to reduce maintenance costs while they are producing. Under optimum conditions they should lay 240 eggs in 12 to 16 months and then weigh 1.8 to 2.0 kg at the end of a laying year.

Laying hens generally trace at least part of their ancestry to the White Leghorn breed. Current commercial strains include Tatum, Schaber, DeKalb and Hi-Line. Such laying types have been bred to eliminate broodiness and therefore are generally incapable of serving as brood hens. Cockerels of laying breeds are inefficient converters of feed into growth of body tissue and should not be purchased for this purpose.

The total number of eggs produced by a hen depends on both the intensity or rate of laying and the length of time that laying takes place (time from first to last egg). Broodiness (induced by the hormone prolactin) blocks the laying process and thus reduces the intensity and total eggs produced. High producing strains of chickens must be free from broodiness, have a short winter moult, be persistent (able to continue laying throughout most of the year) and have the ability to lay long clutches. These characteristics are dependent upon the production of certain hormones, each with its special effects; for example, higher production would be associated with less prolactin, more gonadotropic (growth) hormone, and reduced amounts of thyroxin (which controls the rate and intensity of moult).

Hens that are laying eggs will have larger combs and wattles and a light blue, pink or white vent rather than yellow. Active layers will have pubic bones that are flexible and wide apart; this is necessary to allow passage of the egg.

Moulting is a phenomenon naturally occurring after birds have been in production for 10 to 14 months. It is associated with a cessation of sexual activity. Egg laying stops, feathers are shed in the order of head, neck, body, wing and tail, and the comb becomes pale and smaller. Normally, better layers moult later than the poorer layers. In practice all hens might be force-moulted concurrently when they reach a pre-selected production level. They will then be brought back into production later. Or, because of this tendency to go out of production for a period of time (moult), layers might be butchered after one year and replaced with pullets just coming into production.

Meat Type

Ability to convert feed rapidly into body growth is of the utmost concern in selecting birds for meat production. Body size is of special importance because it places a ceiling on growth. Large strains grow the fastest. Animals reaching slaughter weight earlier will be maintained for a shorter time; this translates into a more efficient feed utilisation. Table poultry with a white colour is preferred because coloured chickens tend to have pigmented pin feathers (those not yet broken through the skin at slaughter time) which detract from the appearance of the carcass.

Meat type breeds rapidly grow large plump bodies weighing up to 4.5 kg. The hens usually lay less than 160 eggs in up to 9 months' lay. Commercial chicks are produced by crossing broad-breasted males with hens selected to grow rapidly. The Cornish breed is usually used in developing table poultry. Dual purpose breeds have lighter weight but higher egg production than the Cornish. They include the New Hampshire, white or barred Plymouth Rock and Rhode Island Red breeds. The Hubbard and Indian River are strains commonly available on the world market.

In developed countries, table poultry is often slaughtered at 6.5 to 8 weeks weighing 1.8 kg; 11% of the weight is lost as blood and feathers and 20% is lost in the viscera; the carcass represents 69% of the original weight of the live chicken. Young, fast growing chicks add weight with the least feed required per unit of weight gained. By growing the chicks to larger weights (e.g. roaster weight of 3 to 5 kg) the same annual yields of food energy and protein can be achieved. However, as the birds age and/or become heavier, about one-quarter more feed is required to put on an additional unit of weight. They also require more space. These disadvantages might be offset by the need for fewer replacements, less care being required for the older birds, and higher meat yield per bird.

BREEDING

Chickens used in large-scale commercial operations are hybrids generally developed and bred by large-scale international poultry-breeding corporations. There are specialists whose breeding flocks produce the eggs that are incubated in hatcheries. These birds will then serve as replacement breeders in multiplier flocks which produce the eggs to be hatched to supply the commercial poultry and egg producing operations. Since these hybrids will not breed true, they might not be as suitable for a small farm breeding flock as the purebreds.

In general, the tropical indigenous bird weighs about 0.9 to 1.8 kg and has a well-fleshed, compact body rather lightly covered with down-free, wiry feathers. Females usually lay up to 3 clutches of 12 to 18 eggs each year, each egg weighing perhaps 28 g. Production can be extended if the eggs are not allowed to accumulate. Under wild conditions the hen stops producing after each clutch and becomes broody, intent on hatching her chicks. These local birds are normally active and vigorous foragers, well adapted to tropical environments.

Exotic purebred chickens have been used to upgrade indigenous village flocks, but this operation is successful only if there is a parallel improvement in feeding, sanitation and management. If exotic cockerels are used, all of the indigenous cocks which are usually more vigorous must be removed. The native cocks will intimidate and possibly kill the exotics.

With widespread availability of speciality bred chicks it is often better to purchase than to breed chicks. However, when the farmer breeds his own replacements, it is useful to know the genetic progress that might be possible. Table 4 shows the relative importance of genetics and environment on several traits.

Table 4   Effects of Environment and Heredity on Traits of Chickens

Trait	Controlled by Environment	Controlled by Inheritance
	%	%
Comb type	none	100
Body type	10–30	70–90
Egg size	50–60	40–50
Egg production	70–85	15–30
Mortality	82–88	12–18
Reproductive disorders	95–97	3–5

Laying hens should be selected for early sexual maturity as determined by the age when they lay their first egg. If light breed birds, such as Leghorns, are to lay 250 or more eggs during the pullet year, they should begin laying when 5 to 5.5 months old. General purpose breeds, such as Rhode Island Reds, should start laying when they are about 5.5 to 6 months old. The best layers will be the first 75% of birds of the same age to come into production. In commercial operations, even though they have higher maintenance requirements, birds of medium body size are often preferred to smaller sizes because they produce larger eggs. Both body size and egg size are highly heritable and thus can be readily changed through selection and breeding. Emphasis should be placed on characteristics that relate to production. Colour and other non-productive traits should be largely ignored. The more characteristics on which selective pressure is applied, the slower will be the rate of progress made on any one. Well-bred replacement chicks can be purchased in most areas of the world; however, if they are not available, chicks can be hatched at home. Care should be exercised to select the desired type, whether for egg or meat production.

Although annual egg production is the primary criterion in selection, the poultry-breeder must also concern himself with body size and general appearance, feathering, rapid growth and liveability as well as egg size, shape, shell texture and colour and egg interior qualities.

FEEDING

Some of the more significant nutrient requirements of chickens are shown in Table 5.

Chickens are omnivores, naturally consuming insects and worms as well as seeds and the vegetative parts of plants. Their diets must supply the 10 essential amino acids (lysine, methionine, threonine, tryptophan, histidine, arginine, leucine, isoleucine, valine and phenylalanine) for adults. Fast growing chicks require dietary glycine as well.

Fermentation in and absorption from the digestive tract of the chicken supplies some but not all of the needed vitamin K and B-complex vitamins. Depending on the dietary ingredients, these and other vitamins (except C) may need to be supplemented. The inclusion of antibiotics in the feed or water depresses bacterial synthesis in the gut, consequently increasing the dietary need for some vitamins. Even though chicken manure is one of the richest natural sources of vitamin B_{12}, this vitamin must be supplied in the diet of chickens that do not have access to their own faeces. Some diets require the addition of choline also, when birds are maintained in cages and are not permitted to consume some of their droppings.

As with other species, only a portion of the total (gross) energy in the feed is used by the bird for productive purposes. The partitioning of this energy is illustrated in the diagram below. Assuming a feed containing 4000 kcal, only approximately 2300 kcal are available for maintenance and body tissue growth and egg production. The heat increment can be used to warm the body in cold weather, but is a detriment in hot weather. As well as environmental factors, the values shown are influenced by composition of the feedstuffs, genetic background and age of the bird.

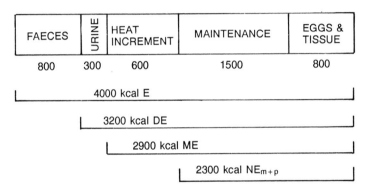

With chickens it is customary to express energy values as metabolisable energy (ME). Digestible energy (DE) caloric values can be generally converted to ME by multiplying DE by 0.82. However this conversion is not as accurate in poultry as in other species; due to the unique digestive system of the bird, digestible energy (DE) calculations overestimate the

Table 5   Nutrient Requirements of Chickens

Class of Chicken	Metabolisable Energy kcal/kg	Crude Protein %	Calcium %	Available Phosphorus %
Replacement chick (pullet) 0 to 8 weeks	2900	18.0	0.9	0.42
Pullet 8 to 20 weeks (maturity)	2900	15.0	0.9	0.42
Meat bird starter 0 to 4 weeks	3200	23.0	1.11	0.55
Meat bird growing–finishing (4 to 8 weeks)	3200	20.5	1.20	0.60
Layer bird	2850	15.0	3.50	0.42

energy value of feedstuffs for poultry. The total energy requirement will depend on the maintenance and egg production costs.

Not only do chickens have a total dietary protein requirement but this protein must also supply all of the 10 essential amino acids. Those amino acids that are most likely to be limiting in a diet containing no animal materials are lysine and the sulphur-bearing amino acids, methionine and cystine. Methionine is said to be the essential one because the body can convert methionine into cystine, but not vice versa. Because of this relationship, often these sulphur-bearing amino acids are combined and considered as a single value. For this reason, in developing rations one should note whether values are for methionine alone or include both methionine and cystine.

The lysine, methionine and cystine requirements of chickens in different production circumstances are presented in Table 6. The amounts of these factors present in various crops are also listed. Maize is noted for its low level of lysine as well as arginine; arginine is required in the diet of growing chicks but not of mature birds. Soybean meal, on the other hand, is the best plant source of lysine. All plant sources are relatively low in the sulphur amino acids (methionine and cystine). A note of caution: all values given in composition tables such as these should be viewed as estimates only. Many factors influence these values such as sampling, analysis, animals used, experimental conditions and procedures, etc.

The protein requirement must cover the protein costs of maintenance as well as the protein deposited in the egg. The efficiency of protein utilisation for maintenance and egg production at a production level of 100% is about 57%; if production falls to 70%, then the efficiency of protein utilisation drops to 46% because an increased proportion of the feed protein is used for body maintenance. Efficiency of utilisation of the entire diet is influenced by the ratio of energy to protein.

With the high performance level expected of the birds, care must be taken that the requirements for protein, including all the essential amino acids, vitamins and minerals, as well as energy, are met. Chickens (as other animals) eat primarily to meet their energy needs. The intake of nutrients, then, is regulated by the energy content in the feed. The ration is developed to provide 2750 to 2900 kcal ME per kg of feed; it is then balanced so that other needed nutrients will be consumed in adequate amounts when that many calories are eaten. Completely balanced commercial feeds are widely available to many small farmers and backyard poultry-keepers. These feeds are formulated assuming a certain intake. Anything that will reduce the feed intake from that assumed in the formulation will require an increase in the concentration of all nutritional factors. Feeds formulated for temperate climates are based on the assumption that some of the energy in the feed will be used to keep the animal warm. This extra feed intake is not necessary in warmer climates. The animal's appetite directs it to eat enough to meet only its energy needs; therefore, in warmer climates, with less energy required, the concentration of nutrients in a feed must be increased. This is because less feed is eaten, but the quantitative requirement for nutrients (amino acids, vitamins, etc.) remains the same

Table 6 Amino Acid and Energy Requirements of the Chicken and Amino Acid Content in Several Feedstuffs

Bird type	Daily Feed[1] g	Protein %	Lysine g	Methionine and Cystine %	Metabolisable Energy kcal/kg
Table poultry 3 to 6 weeks	74	20	14.8	1.00	0.72	3200
Pullets 0 to 6 weeks	21	18	3.8	0.85	0.60	2900
Pullets 6 to 14 weeks	43	15	8.5	0.60	0.50	2900
Laying hens	110	15	16.5	0.64	0.55	2900

Feedstuff	Protein % DM	Lysine % DM	Methionine % DM	Cystine % DM	Energy kcal ME/kg DM	As Fed
Lucerne meal	17.5	0.73	0.23	0.20	1490	1370
Barley grain	11.6	0.40	0.17	0.19	2966	2640
Beans, dry pinto	24.4	1.87	0.25	0.10	2029	2640
Cottonseed meal (41%)	41.4	1.71	0.52	0.64	2666	2400
Cowpea (garbanzo seeds)	26.6	1.99	0.35	—	3360	2990
Fishmeal (herring, mechan. extracted)	72.3	5.70	2.10	0.72	3430	3190
Lupin, white	33.7				2952	2772
Maize, dent, white	11.6	0.29	0.20	0.15	3764	3350
Milk, fresh, whole	27.4	2.27	0.69	—	4954	644
Milk, skim, dried	35.7	2.52	0.96	0.47	2705	2545
Oats, grain	11.4	0.44	0.19	0.21	2865	2550
Peanut seeds w/o shells	29.7	0.97	0.24	0.48	4854	4575
Rice	9.0	0.31	0.18	0.13	2995	2665
Soybean meal (44% protein)	44.0	2.93	0.65	0.69	2506	2230
Soybean seeds (heated)	37.9	2.55	0.47	0.50	3666	3300
Sunflower meal (decorticated solvent)	46.8	1.82	0.77	0.79	2495	2320
Sunflower meal (solvent)	31.8	1.24	0.46	0.42	1714	1543
Sunflower seeds	18.6	0.82	0.43	0.13	3050	2788
Wheat (soft red winter)	13.5	0.38	0.22	0.43	3506	3120

[1] As fed basis assuming 88% dry matter (DM).

and must be supplied in less total feed. Thumb rule: birds consume 1% more feed with each 0.55°C drop in temperature and will eat 1% less with each 0.55°C change in temperature below or above 21°C.

As an example, in the temperate zone hens with a mature weight of 1.8 kg (4 lb) when at maximum egg production will consume 312 to 336 kcal ME daily of rations containing 2750 to 2860 kcal ME/kg and 15.6% crude protein. In warmer climates, the energy requirement of such birds might be 10% less, and they would correspondingly eat 10% less of this feed. Thus, in order to maintain the same level of nutrient intake, it would be necessary to increase the nutrient concentration by 10%. The energy concentration of the ration should remain the same but protein, vitamin and mineral content should be increased by 10%. In a like manner, the addition of fat to a feed increases its energy level, which in turn will decrease the quantity of feed consumed. This presents the need to increase the concentration of other nutrients in the feed.

The bird becomes profitable only after it has consumed more feed than is necessary to maintain its life. Palatability is important. Management practices designed to increase consumption are: frequent stirring of the feed, pelleting the feed, and providing an extra hour of light in both the morning and evening. Growing birds and larger birds, even when laying the same number of eggs, require more feed than grown smaller birds because of requirements for growth and higher maintenance costs of larger birds.

Metabolisable energy to protein ratios (ME/p) are determined by dividing the kcal ME/kg of feed by the percentage of protein in the feed. These ratios vary from 166 to 170 for young growing hens to 196 to 200 for older hens in declining production.

Fast growing chicks need a high level of protein (up to 22%). This dietary need decreases as the growth rate tapers off. The daily feed consumption is influenced by energy and other nutrient concentration in the feed, age and body weight, environmental temperature and level of production.

A deficiency of any one of several of the nutrients often produces the same general signs such as retarded growth, weakness and poor feathering. In the case of proteins, both the total protein and adequate amounts of each of the essential amino acids must be supplied. A deficiency of protein in an otherwise adequate diet results in the increased deposition of body fat. This is because there is not enough protein to make use of the dietary energy that would otherwise be used for growth or egg production. The energy thus becomes surplus and is deposited as fat.

If birds are kept on the ground they will probably get enough minerals from the soil to meet many of their mineral requirements. An exception might be phosphorus and laying hens, especially, will require supplementary calcium. When birds are maintained off the ground, however, special attention is needed to see that the diet provides all essential minerals. Added grit may not be necessary if the feed is finely ground, but as a general rule, grit should be included to help maintain the muscle tone and lining of the gizzard and to grind any coarse or fibrous feeds.

Layers have a high calcium requirement to offset the calcium lost in the egg shell. Even more calcium is needed in layer diets during hot weather to reduce a deterioration in the egg shell quality that is associated with elevated temperatures.

When alterations in the diet are desirable for economic or other reasons, one feedstuff can replace another. Usually, a cereal grain can replace another cereal grain and one protein source can replace a second protein source without significantly changing the ration formulation. Examples would be substituting sorghum grain for maize and degossypolised cottonseed meal and fishmeal for soybean meal. Whenever a protein source is cheaper than cereal grain it can, to some extent, be fed to supply energy. Adding high fibre feeds initially will increase consumption as the bird tries to maintain its energy intake, but when the bird can no longer increase its intake because of gut-fill, additional fibre will reduce feed and nutrient intake and hence egg production.

A single feed to meet general needs might contain 20.5% protein and 2756 kcal ME/kg and consist of 3 ingredients with the composition:

Ingredients	Protein (%)	Energy (kcal/kg)	Proportion in mix (%)
Maize	9.3	3380	54
Soybean meal	40.3	2450	38
Dicalcium phosphate	—	—	2
Limestone	—	—	2
Vitamin–mineral	—	—	2

Additional discussion on feed and feed formulation is in Chicken Learning Activities VI and VII.

Fresh clean water must be available to the birds at all times. In the tropics a hen will consume about 500 ml (1 pint) daily as compared to only 200 to 250 ml in the temperate zone. Lack of water for only a few hours can affect egg production for several days.

Improved feeding of locally bred native breeds is likely to increase their production. If these birds are to be upgraded by crossing with exotic breeds, improved feeding is even more important. Birds scavenging around the family dwelling will probably get their needed vitamins and minerals as well as some protein, but should be fed supplementary grain to meet energy needs. Preferably, the birds should be put in cages off the ground to control internal parasites, in which case they should be fed complete rations. Genetically upgraded birds are likely to have less scavenging ability. Some fresh green feed daily is desirable.

MANAGEMENT

There should be enough feeder space so that at least half the birds can eat at one time.

A laying bird will consume daily about 115 g (0.25 lb) of feed and void about 160 g (0.35 lb) of fresh droppings. To preserve the fertiliser value and reduce odours, manure should be kept dry. If stored wet, as much as 85% of its nitrogen can be lost, primarily as ammonia (NH_3) in as few as 5 days, thus decreasing its fertiliser value. Ammonia up to 20 ppm (mg/litre) will damage the respiratory tract of chickens; 25 to 35 ppm ammonia will irritate the eyes of most people. Dustiness increases the incidence of respiratory diseases and spreads some pathogens.

Crowding and other stress conditions can result in feather picking, cannibalism, mortality and reduced production.

The reproductive (laying) cycle of chickens is influenced strongly by the amount and distribution of light during the 24–hour period. Twelve hours of light are critical. Fewer hours have a negative impact on laying; 14 to 16 hours are usually optimal.

Well-managed pullets can begin laying eggs at 20 weeks of age. Since hens lay 20% to 25% fewer eggs in their second year, layers in commercial flocks are frequently replaced each year. In order to minimise the costs of purchasing or raising replacements, hens might be allowed to moult and be retained. These hens are usually force-moulted in such a way that they all return to production at the same time and are retained for up to 18 months. In farm flocks, hens often moult when they will and are kept for as many years as they produce eggs. Pullet eggs are smaller than those laid by mature hens.

Culling is an ongoing operation. From the very beginning, weak, runty and slow growing pullets should be eliminated. Laying hens will show evidences of production by having a bright red, smooth, glossy, large comb, and a vent that is large, moist and smooth with a bluish-white colour. The pubic bones are thin, pliable and spread apart two or three finger-widths. There is a distance of three or four finger-widths between the pubic bones and the keel bone. As heavy laying progresses, the yellow colour gradually leaves the legs and beak.

Chickens of all ages should be raised in confinement where they and their environment can be better controlled; this is especially true in the tropics. Unless eggs are to be hatched, cocks should not be allowed to run with the hens; this is because there is no need to feed a non-productive rooster and there is no difference in nutritive value between a fertile and an infertile egg.

Replacement of layers

Well-bred replacement chicks can be purchased in most areas of the world making it unnecessary to have sitting hens or incubators; however, care must be exercised to select the desired type. For example, do not purchase layer type chicks to grow for table poultry.

In producing his own replacements the farmer can practise selection either on a flock or an individual basis. He should keep adequate records so that he can select for factors such as: early age at coming into lay, annual egg production above the average of the flock, weight of egg, hatchability, viability of chicks, daily liveweight gain, mature weight, efficiency of feed conversion and adult viability. Cockerels should be the healthiest and fastest growing birds in their hatch.

Clean nests must be available to the hens. Dirty eggs hatch poorly. Up to 82% of all eggs that are set usually hatch. Attention needs to be given to the cock who should be at least 6 to 8 months of age. He can serve 12 to 15 hens. Since cocks do play favourites, some hens might not be served. Eggs laid 24 to 26 hours after mating may be fertile and the fertility drops on eggs laid after 7 days. Fertile eggs remain hatchable for up to 14 days if kept at 10°C to 21°C (30°F to 70°F) and a relative humidity of 60% to 70%. Eggs for incubation should be average size with clean, sound shells. Some diseases such as pullorum (bacillary white diarrhoea) are passed through the egg.

The incubation period for chickens is 21 days, which is the same whether incubation occurs naturally or artificially. Normally, the hen sits on a clutch of 8 to 12 eggs. She should have clean water, a maintenance ration, protection from insect pests and vermin, and, on hatching, protection for her and her chicks from predators.

Artificial incubation requires a constant temperature of 38.5°C to 38.9°C (101°F to 102°F) and 58% relative humidity for the first 18 days, then 70% relative humidity. Strict sanitation must be practised during all phases of incubation. All the equipment should be scrubbed with a disinfectant after every hatch.

Under natural conditions the broody hen protects and looks after her chicks. Artificially incubated chicks require the careful attention of the manager. The chick is born with a stomach filled with yolk material and therefore needs no food for the first day. Chicks are to be kept warm and dry without crowding or overheating. Chicks purchased from commercial hatcheries should be kept separated from other chickens, primarily to control diseases. Chicks for egg production should be sexed at the hatchery and only female chicks should be raised.

At 6 to 8 weeks the cockerels should be separated from the pullets and all birds transferred from the brooder to a rearing unit. Only birds that are of the same sex and size should be kept together. Range rearing permits birds to scavenge for herbage, seeds and insects but requires supplementary feeding, protection from predators and a protective enclosure for the night. Being on the ground exposes the birds to internal parasites. Such birds should be moved to a different area at regular intervals to avoid buildup of internal parasites and disease organisms.

Environmental Control

Chickens, like many other animals, are better adapted to keeping warm than keeping cool. The zone of thermoneutrality for adult egg-laying birds is 20°C to 30°C, depending on the relative humidity. Exposure to temperatures down to 7°C increases feed consumption. Colder temperatures decrease egg production, and at −10°C combs and wattles freeze and hens stop laying. As the temperature rises above the thermoneutral range the birds decrease activity and feed consumption and the respiration rate increases with open mouth panting. Since birds do not have sweat glands, they dissipate heat by evaporation through the respiratory system. Using this evaporation for cooling increases the water requirement. High humidity increases the stressful effects of high temperature because the moisture covering the mucous membranes does not evaporate as readily. Air movement slightly increases comfort. Birds decrease their feed intake in order to reduce internal heat production. High ambient temperatures may initiate a partial moult and drastic drop in egg laying. If exposed over time, however, egg production might be expected to improve, indicating a considerable degree of acclimatisation. Eggs laid during heat stress are often soft shelled and misshapen; this is because the rapid breathing of a bird in hot weather hyperventilates the lungs, reducing the carbon dioxide (CO_2) available in the blood reaching the uterus or egg shell organ. This CO_2 is converted to carbonate (CO_3) which is combined with calcium (Ca) to form calcium carbonate ($CaCO_3$) for the egg shell.

Keeping chickens up off the ground in cages can reduce infestation with worms and other internal parasites and diseases. Individual cages are desirable but not necessary. Cages (pens) can be made from local materials such as bamboo, welded wire or other strong material with side and top openings about 2.5 × 5 cm with smaller openings in the floor. The objective of the pens is to keep the bird comfortable but confined. There are any number of ways in which this can be done. Both table poultry and layers can be raised in cardboard dry goods boxes bedded with shavings, sawdust, straw or shredded newspapers. Tin cans or bamboo joints tied to the side of the cage can serve as waterers and feeders. Screen, hardware cloth or woven bamboo can be used to cover the boxes, allowing adequate ventilation while preventing escape. Where several birds are raised together, each chicken should have a minimum of 450 sq cm of floor space, and adequate feed space should be provided so that at least half of the birds can eat at one time.

Housing for poultry should provide shelter from the elements, roosting facilities, protection from predators and exposure to wild animals, a healthy, comfortable environment and should require minimum labour by the operator. Buildings should be located in a well-drained area. Depending on the weather, walls may not be needed but sides should be screened to keep out larger vermin and wild birds. Shade and adequate ventilation are effective in reducing heat stress. Minimal space allotment per bird in laying houses in cages should be 0.05 to 0.06 sq m (0.5 to 0.7 sq ft) and 0.18 to 0.28 sq m (2.0 to 3.0 sq ft) if birds are to be kept on the floor with either built-up litter or slotted floor systems. Up to 3 laying hens can be kept in a cage 30.5 × 45.7 cm and 35 to 46 cm high without hampering their production. However, these birds should be debeaked to prevent the cannibalism that often develops when birds are so confined.

HEALTH AND DISEASE

Poultry that have been well fed and managed, including vaccination against local diseases, usually remain healthy especially if they are few in number and given ample space. Close day-to-day observation of the flock is the only way to attack disease problems before they get out of hand. General symptoms of disease are coughing, sneezing, watery eyes, laboured breathing, droopiness, rough feathers, bloody droppings, and a sudden drop in feed consumption and egg laying.

Ascites is a condition in which serum-like fluid escapes into and accumulates in the abdominal cavity; it is sometimes referred to as 'water belly'. There may also be a venous congestion elsewhere in the body. It is most common in male table poultry. Although it is most frequently found in individuals, entire flocks can be affected.

Disease prevention is more important and economical than treatment. In a disease outbreak, sick birds should be separated from healthy birds and strict sanitary measures applied. Very sick birds should be killed and dead birds should be necropsied, burned, or buried deeply enough so that they will not be dug up by scavenging animals and the diseases spread. Animals of different ages running together tend to perpetuate and compound disease incidence. Vaccination for many diseases is available, but only healthy birds should be vaccinated.

There are many diseases that have been identified. Some are more important in some localities than others. The flock owner must learn to recognise and control those important in his own area. Some diseases of particular concern in many areas include:

Fowl typhoid is caused by *Shigella gallinarum* which is closely related to *Salmonella pullorum*. These are bacteria that can survive for 2 months in deep litter. The disease can be spread to healthy animals by infected birds that do not show signs of the disease, by shoes, clothes, feed, water, rodents or wild birds. Symptoms include a wasting away, weakness, drowsiness, yellowing, diarrhoea, and anaemia. Sometimes death occurs before other symptoms appear. The comb and wattles are pale after death. A vaccine is available.

Fowl cholera is caused by a virulent *Pasteurella avicida* bacteria. Some forms of this microorganism can attack man and other animals. It can be spread through contaminated feed and water; organisms are carried by flies and rodents, and in the dust and droppings of infected birds. Dead birds can infect others for at least a month after death. The combs and wattles of these dead birds are dark in colour. The mortality rate is very high. Birds that appear to have recovered may still carry the disease. The addition of tetracycline to the drinking water is helpful in controlling the spread of infection. Symptoms include watery diarrhoea and difficult breathing, with death sometimes occurring before other symptoms appear.

Newcastle disease, known also as avian pneumoencephalitis, is caused by a virus that is unusually resistant to heat. It can spread directly or indirectly from excretions, improper carcass disposal, wild birds, air movement around infected birds and contaminated equipment. The incubator can be a source of infection. Symptoms include respiratory problems such as gasping, coughing and sneezing, and hoarse chirping. Birds become inactive, stop eating and laying eggs. Later nervous symptoms such as paralysis and muscular tremors appear. Death losses are high, especially in young birds. Vaccines are available.

Fowl pox occurs most commonly when young birds are put in an infected laying house. A virus enters the bird through wounds and abrasions on the

101

head or in the mouth, or when carried by mosquitos. The disease spreads slowly but may persist on the farm for years. The disease can be controlled by not overcrowding, by avoiding wounds and injuries and by following sanitation and vaccination procedures.

Botulism (limberneck) is caused by ingestion of the toxin elaborated within the cell walls of the spore-forming *Clostridium botulinum* organisms. Birds eat feed contaminated with this organism such as in decomposing dead animals, spoiled canned food and drinking water from stagnant pools of water. This is the most deadly toxin known.

Bumblefoot is present more often in older, heavier birds. It is an abscess (*Staphylococcus aureus* infection) of the foot caused by bruising or cuts getting infected, with caseous (cheesy) material developing on the ventral surface of the foot, requiring surgical and medical treatment.

Additional Diseases

The following additional diseases of chickens are largely under present-day control and eradication in developed areas, but they may pose a threat in certain other areas of the world. They need to be constantly guarded against.

Pullorum disease (bacillary white diarrhoea) is one of the most serious diseases of times past. In 1900 Rettger discovered it to be caused by *Salmonella pullorum bacillus*. Infected baby chicks have severe white diarrhoea and a very high mortality rate, and survivors carry and transmit the disease. Pullorum was largely eradicated in the 1920s and 1930s through blood testing of breeding flocks. The organism is transmitted through the eggs from infected hens to the baby chicks. Thus, the most important preventive measures are:
  a. blood testing breeding flocks;
  b. developing birds with high degrees of natural immunity.
Baby chicks should be purchased only from breeding flocks and hatcheries that have certified blood-tested Pullorum-free chickens.

Other salmonella (*Paratyphoid*) infections caused by the large family of pathogenic salmonella organisms are always a threat in poultry raising, requiring constant vigilance and careful sanitation. *S. typhimurium, S. enteriditis, S. meleagridis, S. california, S. newport, S. minnesota, S. illinois, S. suipestifer* and many others are potential pathogens.

Fowl tuberculosis, caused by *Micobacterium avium*, is another disease that has largely been eradicated in many areas of the world. It always poses a threat and must be constantly guarded against. It is not only contagious to birds but also to rabbits and pigs. Man is highly resistant, but in a few instances he, too, has been infected. Fowl tuberculosis causes chronic disease lesions throughout the body, especially in the liver, spleen, intestine and bone marrow. Testing and sanitation are the keys to control.

Infected flocks must be slaughtered and environs carefully cleaned and sanitised.

Infectious fowl roup (sinusitis—head colds) is caused by *Haemophilus gallinarum*. Outbreaks usually follow a stressful condition such as changing housing or other environmental circumstances, especially during inclement weather. Roup is characterised by swellings of the head, noisy breathing, abscesses of sinuses and head cold-like symptoms due to greyish yellow exudate on the respiratory mucous surfaces. The mortality rate is low, but loss of egg production follows for days and weeks. Note: nutritional roup is a consequence of a vitamin A deficiency and exhibits symptoms similar to those developing in a vitamin A shortage.

Laryngotracheitis infection is a viral disease which can be controlled by vaccination. The condition is characterised by severe cold-like symptoms including laryngitis, tracheitis, open-mouth breathing, coughing, yellow and cheesy discharge from the mouth and throat. The illness lasts 7 to 15 days (average), with a mortality rate of 5% to 60% (average of 13%). Since survivors may be carriers, follow-up vaccinations and a careful hygiene programme must be followed to eradicate the disease.

Fowl plague (fowl pest) is an extremely serious, contagious viral disease. Fowl plague is very highly fatal, having symptoms like fowl cholera, affecting chickens, turkeys, geese, pheasants, some water fowl, and many wild species of birds. Pigeons, ducks and some other water fowl have more resistance than do chickens. Cardinal signs of the disease are extremely acute infection with very high and rapidly fatal death rate, and general haemorrhagic conditions of the body found in autopsy. If suspected the disease should be reported immediately to the proper veterinary regulatory officials for testing, diagnosis and eradication. Eradication requires that all diseased and dead birds that have been exposed must be burned and stringent cleaning and disinfection of premises and equipment be performed.

Fowl paralysis (range paralysis, fowl leukaemia, leukosis or Marek's disease). Blood tests, eradication of carriers and the selection of highly resistant breeding stock have largely eliminated this serious viral disease over the past 50 to 60 years. It is characterised by abnormally high lymphocyte blood cell counts, lymphoid tumours in the visceral organs, skin and muscle. There is a gross enlargement of the nerves, and other organs are affected as well.

Internal Parasites

Coccidiosis is caused by a protozoan, a microscopic parasite that is widespread in the soil and in adult birds that have developed an immunity. Caecal or bloody coccidiosis is most commonly seen in 5 to 7 week old chickens. The resistant stage of the organism (oocyst) is passed in the droppings and survives in the ground for long periods. These oocysts can contaminate feed and water, which then infect the birds. The disease can be prevented by raising young birds off the ground or on ground that has not been exposed to older birds for at least a year, and by separating older birds from their own faeces. Symptoms include bloody droppings, decreased feed consumption, droopy appearance, weight loss and high death losses. The addition of amproline or sulphonamides to the feed mixture is effective in controlling the disease if treatment is started early in the lives of the birds.

Worms   Over 60 species of worms occur in poultry. Small farm flocks that are on the ground, either freely ranging or in back yard pens, are generally infested with one or more of these parasites. Internal parasites commonly encountered are nematodes (roundworms) that include the large roundworm, caecal worm, capillaria (hair) worm, and several species of tapeworms and flukes. The roundworms are elongated, cylindrical and unsegmented. They have a well-developed digestive tract, and, in contrast to the tapeworm, are either male or female. Their bodies are covered with a tough layer, called a cuticle, which protects them from the digestive juices of the gastro-intestinal tract. These worms produce microscopic eggs which the host voids in the faeces.

1. To become infested with **roundworms**, a chicken must first eat the infective eggs. Ascarid (a genus of nematode) infestation is made directly from one animal to another through worm eggs in the faeces, whereas most other nematodes involve an intermediate host such as earthworms, grasshoppers, beetles, cockroaches and sowbugs (woodlice).
2. **Flatworms** (tapeworms) are flattened, ribbon-shaped and usually segmented. They have no digestive system but absorb feed from the chicken digestive tract through their body surfaces. Worm eggs and worm segments are voided in the faeces. Tapeworms require an intermediate host for the completion of their life cycles. Common intermediate hosts are house flies, dragon flies, fish, slugs and snails. The chicken becomes infested by eating the infective intermediate host. Confinement rearing in cages off the ground can eliminate contact with the faeces and intermediary hosts.

Treatment for internal parasites consists of the use of one of several broad spectrum antihelmintics such as coumaphos, levamisole, phenothiazine and thiabendazole. However, drugs alone cannot produce effective worm control. Drugs might reduce the level of contamination, but concurrent management changes must be made. In fact management is the only control available for some of the flatworms.

Prevention and control of internal parasites depends on sanitation and management practices including:

   a. Keeping the birds separated from their faeces;
   b. Thorough cleaning of and disinfecting the premises between groups of birds;
   c. Separation of young from old birds;
   d. Control of free-flying birds and of insects and other forms of life that serve as intermediate hosts.

Symptoms of infestation include diarrhoea, general unthriftiness and listlessness, ruffled feathers, loss of weight and production. Slight infestation is not easily detected, but heavy infestation can be detected on postmortem examination, and can be the cause of death.

External parasites

Lice are flat, wingless, brownish yellow insects that move very quickly. They are the most prevalent external parasite on chickens and consist of the common large louse, small body (shaft) louse, chicken head louse and wing louse. With their chewing and biting mouthpieces they consume cells of the skin and feathers and even blood. They spend their entire lifetime on the chicken and cannot live long without food.

Mites are smaller than lice and are not readily visible with the unaided eye. There are many types with different body shapes and life styles. Since most mites use blood or lymph as food, they can cause anaemia. These blood suckers can transmit diseases such as fowl cholera, Newcastle disease and several types of encephalitis. Some mites live their entire life on the bird; others can live in an empty chicken house for months without food, lodged in cracks, crevices and undersides of roosts.

Other insects such as chiggers, bedbugs, flies, ticks, gnats and mosquitos can also cause injury to poultry. Careful examination especially around the vent can usually detect any infestation of external parasites. Pests irritate the skin and often cause scabbing. The irritation and restlessness increase the bird's activity and decrease its feed intake. The first result is a loss in body weight and a reduced egg production.

Treatment of external parasites with one ounce of malathion in a gallon of water (7.5 ml/l) sprayed on the birds and 15 ml/l (2 oz/gal) sprayed on the roosts repeated at 10-day intervals is usually effective. Dusting birds and litter with an appropriate insecticide can also be effective.

Insecticides are poisonous to man and animals as well as insects. Read the labels on the containers carefully and follow the directions exactly. Do not contaminate the feed, water or eggs.

As with any toxic materials, drugs and insecticides must be handled with caution, taking care that the manufacturer's directions are followed. These materials and the containers in which they come should be protected from accidental encounter with animals and children, and proper disposal needs to be made of the empty containers.

Animals that die for no recognisable reason should be opened and examined to learn the cause of death.

TURKEYS

The turkey of today originated from the wild birds native to North American forests. They were domesticated by Indians in pre-Columbian times. Probably, the current domestic varieties developed largely from those birds that were taken by explorers to Europe where they were improved and then spread to many parts of the world including the Americas.

In general the same principles of breeding, brooding, rearing, feeding and management apply to all species of birds, but there are differences, some more important than others.

Reproduction

Turkeys might begin mating at 30 weeks of age although they do not reach full maturity until a year later. They are season breeders, mating when days are growing longer. In temperate climates under good management, hens of improved varieties might annually lay up to 90 eggs weighing 85 g (3 oz), whereas in the tropics those birds of native breeding might produce only 20 eggs weighing 57 g (2 oz).

Toms of heavy breeds can mate with up to 10 hens, 14 hens for small-type breeds. Due to the large size of the tom in relation to the hen this mating can damage the hen especially if the tom's spurs are long and sharp. This may not be such a problem in smaller native varieties. Where this is a problem, some have resorted to covering the hen's back with a cloth material or using artificial insemination.

Hens like roomy, well-protected nests; 28 days are required for incubation.

Feeding

Poults are larger than chickens, requiring more space in the brooder and at the feed and water troughs. It may be necessary to teach artificially hatched young turkeys to eat and drink by placing their beaks in the feed and water. Turkeys are creatures of habit and have an aversion to changes in their feed or routine.

The protein content of turkey feed should be about 5% higher than that for chickens. Starter rations should have 26 to 28%, grower rations 20 to 23% and finishing rations 14 to 16% crude protein. Mature turkeys can tolerate more fibre in their feed than can chickens; 5% fresh grass can be included in the diets of adult turkeys.

Under usually encountered conditions of climate and good management an 11 kg tom can be produced in 30 weeks requiring 4.5 kg of feed/kg gain. Under optimal conditions the same bird can be produced in 20 weeks at a feed cost of 3.1 kg/kg gain. About 70% of the production cost is the cost of feed.

Management

The wild turkey is a hardy bird but some of the survival qualities have been lost in the development of the meatier, broader-breasted modern varieties.

Turkeys in developing countries are generally allowed to range. This practice may lessen feed costs, but profitability is reduced because of losses from internal parasites and soil-borne diseases, thievery, predatory animals and adverse weather conditions. On the other hand the disadvantages of confinement rearing include increased feeding, equipment and housing costs, increased respiratory diseases and cannibalism. However, these disadvantages do not usually offset the advantages of a confinement over a ranging managing system.

Adult turkeys can withstand cold and rain, but they are sensitive to wind and heat, especially when accompanied by high humidity.

Turkeys should have water available at all times. If they are given free access to water after being denied water for an extended period, they will drink excessively, become water intoxicated, develop a severe diarrhoea and possibly die.

Each day birds should be inspected, giving attention to the eyes, nostrils and corners of the mouth, the feathering especially around the vent, signs of cannibalistic attacks and diarrhoea, and the general appearance and behaviour.

Turkeys are especially susceptible to blackhead, coccidiosis and erysipelas. Blackhead is a protozoal disease caused by *Histomonas meleagridis*, together with a bacterial infection. It is especially serious in 8 to 16 week old poults. Both the liver and caeca are involved. Common symptoms are lethargy, yellow diarrhoea, a dark (cyanotic) head and death. Since the disease is transmitted by droppings, poults can best be managed on slated or wire floors. Drugs such as furazolidone can effectively control the disease.

Coccidiosis is caused by the protozoa *Eimeria* and is almost universally present in the droppings in poultry-raising operations. Coccidia among birds up to 8 weeks of age become a problem when large numbers of oocysts are eaten. Symptoms include severe diarrhoea (with no blood), droopiness, loss of appetite and weight. Control lies in separating the birds from their faeces and administering any of a wide number of coccidiostats such as sulphonamides.

Erysipelas (also in sheep and pigs) is caused by *Erysipelothrix insidiosa* usually from contaminated soil. Affected birds (toms more than hens) develop diffuse haemorrhages in the large muscles and visceral organs. They lose their appetite, become listless and have a greenish yellow diarrhoea. They may also have laboured breathing, a nasal discharge and discoloured face. Erysipelas can be treated by injecting 11,000 mcg penicillin per kg body weight and an *E. insidiosa* bacterin.

Fowl cholera due to *Pasteurella multocida* is a widespread contagious disease occurring in both acute and chronic forms. Dead birds may be the first indication of the disease. Usually seen are septicaemia, fever, depression, anorexia, mucous discharges from the mouth, ruffled feathers, diarrhoea and pneumonia with increased breathing rate. Hyperaemia with subsurface bleeding is seen in the internal organs. The chronic form is generally related to localised infections and swellings in the sternal bursas, wattles, joints and foot pads. Exudative conjunctivitis and pharyngitis and/or wry neck may occur.

Prevention of the disease depends on good management. Cold, damp weather favours the disease. Poor appetite reduces the effectiveness of the usual treatment with antibiotics in the feed and/or water.

Turkey coryza occurs sporadically worldwide as a respiratory disease.

The causal agent, *Bordetella avium,* is highly contagious and remains in an area for long periods of time. Sneezing and a mucoid discharge are usually the first symptoms seen. These may occur at any age but are more likely in the young at 7 to 20 weeks of age. As the disease progresses the birds become depressed and severe rhinotracheitis with difficult breathing that produces unusual sounds within the body (rales) occurs. A vaccine has been recently developed for coryza control although the best preventive measure is good management practices that reduce stress on the birds.

Infectious synovitis (infectious arthritis) is a crippling disease with a yellow viscous exudate in the swollen sore hock, wing and other joints, the bursa and keel. Several infective agents might cause the disease. *Mycoplasma synovia* can be controlled by tetracycline antibiotics; the level of damage caused by staphylococcosis can be reduced by a vaccine.

Turkeys are prone to panic and stampede, injuring themselves by running into walls or fences, and crowding into corners and becoming smothered.

They can be driven from one place to another. In semi-darkness or darkness the birds can be picked up with both legs without undue disturbance.

PIGEONS

Man's use of the pigeon dates from antiquity. There are hundreds of breeds with various specialities. The most preferred meat is from squabs just ready to leave the nest at 25 to 30 days and weighing 300 to 450 g. They will be completely feathered but have not yet begun to fly. Pigeons may live 12 years and weigh 800 g when mature. Each year a pair should produce 12 to 15 squabs with broad, thick and solid bodies.

Mating behaviour is the easiest way to differentiate the males from the females. The mature male is usually larger with a thicker neck and more rugged appearance. He will coo and strut, spreading his tail feathers, often making complete circles. He drives the female to stay at the task when building a nest and laying the eggs. When billing, the female places her beak within that of the male.

Pigeons are gregarious animals tending to flock together. They tend to pair up for life and each shares in the duties of rearing young. After mating, the female lays a clutch of 2 eggs. During the middle of the day she is relieved from sitting by the male. Eighteen days are usually required for incubation.

The young are hatched naked. They elevate their heads on outstretched necks and give their characteristic squeak for food. This sound stimulates both the male and female to produce crop or pigeon milk. This is a secretion from the crop to nourish the young. The parent opens its mouth, regurgitates and the squab puts its beak far into the mouth of the adult and eats.

At 4 to 5 weeks of age the squab's feathers have filled out and the bird leaves the nest even if it has not yet learned to fly. It is particularly vulnerable to predation at this time. Sexual maturity is achieved when about 4 months of age for the male and 6 months for the female.

DUCKS

Ducks are extremely versatile, thriving under many different conditions of climate, feed and management. Domestic ducks when compared with chickens require more feed, have a stronger flavoured meat and eggs, lay more and larger eggs than indigenous fowls, grow to as great a size, are more hardy, have lower housing requirements, are more resistant to disease and parasites (such as coccidiosis and cholera although they can transmit some of these diseases to chickens), they protect themselves better from predators and they are better scavengers. Once grown, ducks can, if necessary, subsist on grasses, weeds and insects. They eat many young, growing plants so they cannot be allowed in a growing garden; they might, however, keep the slugs and snails and weeds out of a mature garden. A few handsful of grain fed in the evening will keep them nearer the house.

Duck breeds include:

1. The Khaki Campbell that under good management begins laying at 4½ months and can produce yearly 300 to 350 eggs weighing 60 g each. On the average, this level of production is maintained for 3 or 4 years.
2. The White Peking, Aylesbury and Muscovy are meat types reaching 3 kg or more at 7 weeks of age.
3. Mallards (the males have green heads) have been domesticated, but they are smaller than other breeds.
4. The Muscovy differs from the other breeds having originated in South America. They should be marketed before 17 weeks of age. Adult females weigh 3.2 kg and drakes 4.5 kg. The hens are poor egg producers but good sitters. They will hatch and care for approximately 30 ducklings from the 40 to 45 eggs they produce in a year. Other breeds are not good sitters; broody chickens can often be better used. A duck will cover 12 eggs, a goose 9 to 10 eggs; a chicken can cover 4 to 6 goose eggs or 10 to 12 duck eggs. Goose eggs under a chicken should be turned by hand twice daily; they are too large for the hen to manoeuvre.

One drake can service up to 6 females. Ducks prefer to nest as near the floor as possible. Most eggs are laid at night or early morning. Twenty-eight days are required for hatching except for Muscovites, which require 35 days. Dry, clean nests should be provided for sitting ducks and hens. Both feed and water should be conveniently located. Water fowl eggs hatch best when the nest is on moist ground. If the eggs are to be incubated artificially the temperature should be 0.6°C lower than for chicken eggs and the relative humidity never below 60%. From the second to fourteenth day the eggs should be marked and turned twice daily and sprinkled with warm water at least once a day. After the fourteenth day the eggs should be turned and sprinkled 3 or more times per day.

Ducklings need to be kept warm and dry for the first few weeks, but they are not as sensitive to temperature changes as chicks. They should be protected from cold and draught for the first 2 to 4 weeks. Litter on floors should be dry and mould free. Ducklings need shelter from the sun, especially after eating.

Feed that has been pelleted is preferred to mash in feeders. Drinking water should be available at all times. Water for swimming is not necessary.

After 6 weeks of age ducks develop vocal differences. Females have a shrill quack while drakes have a soft voice. Mature drakes develop several curled tail feathers as well as having brighter colours.

Avoid catching and carrying ducks by the legs, which are relatively fragile. With one hand grasp the bird around the neck close to the body. Slide the other hand palm up under the breast to the abdomen, supporting the bird along its length upon the forearm and hand.

Under commercial conditions ducklings other than Muscovy are ready for market when 7 to 9 weeks of age. They should weigh 2.7 to 3.2 kg, having consumed 9.1 to 10 kg of feed.

A duck can be slaughtered by hanging it by its feet and then severing the left jugular vein and carotid artery by cutting the throat on the left side at the base of the beak. After bleeding, grasp the bill with one hand and the legs with the other and submerge the body breast upward in water 57 to 63°C for 1½ to 2 minutes. Adding some detergent and moving the body back and forth will get the water to the base of the feathers. Strip the feathers by rubbing against the lay of the feathers. Eviscerate the carcass and keep cool until ready for cooking.

GEESE

Domestic geese are large, aggressive, noisy, disease resistant, quick growing birds that need little attention when grown and may live as long as 30 years. They are raised for their meat, large eggs, feathers and sometimes to weed crops or to act as 'watch dogs'. There are at least 9 breeds of which the Chinese and African probably do best in the tropics and the Toulouse and Embden in temperate zones.

Since they mate rather selectively and for life, flocks should have 1 gander (male) for every 2 females.

Geese like to swim, but it is not necessary. Goslings do not require brooders but they should be kept warm, dry and free from draughts for the first few weeks. Goslings need larger feeders and waterers than chicks. On commercial operations it is not uncommon for growing geese to weigh 4.5 kg in 10 weeks.

The starter feed should have 20% to 22% protein and preferably be pelleted. Beginning at 6 to 8 weeks of age, geese are selective grazers; they will not eat tough, dry grasses and some other plants. Grazing birds need water available and supplementary grain fed in the evening if sufficient young grasses and weeds are not available.

Geese, like ducks, can suffer broken or disjointed legs if caught and held by the legs. Rather, with one hand grasped firmly around the neck close to the body, slide the other hand under the breast and abdomen. The operator should pinion the wings between the arm and body to avoid painful blows.

As with ducks, geese can be killed by severing the left jugular vein and carotid artery and either dry plucked or scalded before plucking. Pin feathers can be pulled with tweezers or by catching the pins between the thumb and a dull knife.

GUINEA FOWL

Guinea fowl originated in Africa where today they remain in the wild, especially in the drier areas of East Africa. Since the days of ancient Greece and Rome they have been kept for their tasty meat and edible eggs. Guineas are smaller than chickens. They are usually consumed when 16 to 18 weeks old weighing 1.25 to 1.5 kg. A well-managed hen might lay up to 100 eggs a year, each weighing 40 to 50 g.

There are 3 well-known domestic varieties, the Pearl, White and Lavender, the Pearl being the most numerous.

Guinea fowl are normally kept as scavengers. Because they are wilder and do not take confinement well, they are more difficult to raise profitably than chickens.

The birds are characterised by a harsh cry which becomes louder and agitated when they are upset by strangers or other unusual circumstances.

In confinement 1 male can serve up to 5 hens. Hens like to hide their nests; if these nests can be found all but 1 or 2 eggs can be removed at a time. Hens should not be disturbed while laying.

Eggs can be incubated or brooded either by the guinea hen or a broody chicken. Artificial incubation should be at 39.4°C (with no fan) for the first 3 weeks and 40°C for the last week. Brooding temperature should be 40°C the first week, being reduced 2.8°C each week, thereafter.

Guinea keats (chicks) should have a starter feed containing 25% protein for the first 6 weeks after which the protein could drop to 15%. All fowl should be fed in the late afternoon so they will return to the shelter at night.

8
CHICKEN LEARNING ACTIVITIES

I BUILDINGS AND EQUIPMENT FOR CHICKENS

Purpose Learn the structure and functions of the buildings and equipment needed for successful poultry production.

The buildings should exclude cats, dogs, rodents, snakes and numerous wild animals and birds that can frighten and kill chickens, especially the young. Such extraneous animals can also spread diseases and parasites as well as consume and/or contaminate large amounts of poultry feed. Many such animals can also be destructive to buildings by their burrowing and gnawing.

Chickens should be confined in order to control them and their diseases. Confinement also makes easier the gathering of the eggs. If chickens are not kept off the ground, they should be on clean, dry litter to prevent infestation with worms and other internal parasites.

The chicken house should provide:

1. Comfort including protection from sun and rain, hot and cold temperature extremes, wild birds, pests, predators and poachers.
2. Light and ventilation with an absence of draughts.
3. Nests or other provision for laying eggs in a laying house.
4. Facilities for feeding and watering.
5. Protection from disease by being up off the ground or on clean, dry litter.

Building a Cage

Cages for layers: 1, 2, or 3 pullets can be kept in a 30 × 45 cm cage although the performance of each hen is decreased by about 2.5% with each additional bird; the profitability is increased by adding up to 3 hens per cage. Make the sides 45 cm high. The floor should slope downward to the front dropping 1 cm/12 cm. Provide a 5 cm opening so that any eggs will roll down to the front away from the feet of the hen, and where they can be easily gathered (see Figure 1).

Cages for table poultry: table poultry can be kept in colony cages. For the first 3 weeks each chick requires 465 sq cm of space, and thereafter 930 sq cm of space is needed. The same construction principles can be used as for layers except that the floor should not slope because there is no need for egg collection facilities (see Figure 1, p. 7).

Materials

Cages can be built of locally available materials such as bamboo, welded wire or other strong material. Materials used can have openings of about 2.5 × 5 cm in the sides and top and floor openings of 2.5 × 2.5 cm.

Feeders and waterers must be secured to the side of the cage where they cannot be overturned; they can be made of such things as tin cans or bamboo joints.

Both table poultry and layers can be raised in enclosures such as large cardboard dry goods boxes. Shavings, sawdust, straw or shredded newspapers can serve as bedding. Boxes can be covered with material such as hardware cloth or woven bamboo to allow adequate ventilation while keeping the birds in the boxes. Care must be taken to protect the birds from extremes in weather, especially heat, dampness and cold draughts.

Check your learning Can you:

- Explain why chickens should be in cages off the ground.
- Design and build a cage for layers using locally available materials. Provide for comfort and ease of feeding, watering and egg collection.
- Design and build a cage for table poultry.

II SELECTING BIRDS FOR PRODUCTION

Purpose

1. Decide the type of operation, whether for eggs or meat.
2. Decide the type of chicken to best fit the choice.
3. Learn where these chicks can be obtained.

Before acquiring any birds, a management commitment needs to be made. Birds must be fed, watered and otherwise cared for every day of the year, regardless of the type.

Choose the type of poultry operation and the type of birds most suitable to that operation. Before beginning a poultry project it should be decided whether eggs or meat are to be the first consideration or whether a dual purpose bird is preferred. The greatest yields are available from varieties of birds that are specialised either to produce eggs or meat. Table 3 presents approximate data useful in making this decision. It will be observed that under conditions that are only two-thirds optimal, laying hens if kept for 2 years would produce 14,940 kcal of energy and 1140 g of protein on an annual basis. On the other hand, table poultry optimally could grow to 1.81 kg in 8 weeks, then, if each bird were replaced immediately with another chick, there would be a total of 6.5 chickens grown per year (52 weeks ÷ 8 = 6.5). Assuming again an actual production of only two-thirds optimal, these 6.5 chickens would produce 9100 kcal of food energy and 690 g protein each year. If food production were the only consideration, then laying hens should be selected. Economic, personal or other considerations could alter this choice.

Chickens with the best genetic makeup for either egg or meat production are generally obtained by crossing special breeds and strains of birds. These have been scientifically developed by commercial breeders through rigorous selection from large numbers of individual birds. This process utilises hybrid vigour to maximise production. Chicks available to the individual farmer are hybrids that will not breed true; that is, it should not be expected that the offspring of special strains will have the same qualities as the original stock. Commercially produced baby chicks are available in many markets throughout the world at reasonable prices. For these and other reasons, it is often more profitable to purchase replacement chicks rather than trying to produce them on the farm.

Although older hens lay larger eggs, they produce 20% to 25% less eggs in their second than in their first year.

Chickens for egg production should have small bodies to reduce maintenance costs while they are producing. Under optimum conditions they should lay 240 eggs in 12 to 16 months and then weigh 1.8 to 2 kg at the end of a year's laying.

White laying hens generally trace at least part of their ancestry to the White Leghorn breed. Current commercial strains have been bred to eliminate broodiness and therefore are generally incapable of serving as brood hens. Cockerels from laying type breeds are inefficient converters of feed into growth of body tissue and should not be purchased.

Meat (table poultry) type breeds rapidly grow large plump bodies weighing up to 4.5 kg. The hens lay less than 160 eggs in up to 9 months' lay. Commercial chicks are produced by crossing broad-breasted males with hens selected to grow rapidly. The Cornish breed is usually used in developing table poultry. Dual purpose breeds are a lighter weight but have higher egg production than the Cornish. They include the New Hampshire, White or Barred Plymouth Rock and Rhode Island Red breeds. The Hubbard and Indian River are strains commonly available on the world market.

In commercial operations, table poultry is often slaughtered weighing 1.59 kg at 7 to 8 weeks. 11% of the weight is lost as blood and feathers and 20% is lost in the viscera, the carcass representing 69% of the original weight of the live chicken. Young, fast growing chicks add weight with the least feed required per unit of weight gained. By growing the chicks to larger weights (e.g. roaster weight of 2.72 to 3.62 kg) the same annual yields of food energy and protein can be achieved. However, as the birds age and/or become heavier, about a quarter more feed is required to put on an additional unit of weight. They also require more space. These disadvantages might be offset by the need for fewer replacements, and less care is required for older birds.

Native breeds as traditionally managed are usually not confined and reproduce on the basis of natural selection. They usually seek their own feed, supplemented with a little grain and table scraps from the kitchen. They are likely to be infested with parasites, to be small and to lay very few eggs, often in difficult places to locate.

Investigate the local market to learn if and where improved varieties of chicks can be obtained. Check the local markets, visit any local commercial hatcheries or poultry producers, and contact any importers to learn availability and prices of the type of chicken selected. Unless the chicks have been sexed, 50 chicks should be purchased to get 23 to 24 pullets. Raising the cockerels of laying varieties for meat is inefficient and not recommended.

If no chicks are available for purchase, consider the feasibility of obtaining fertile eggs from improved varieties and artificially incubating them or putting them under broody hens for hatching. All chickens should be kept off the ground and separated from their own faeces to prevent infestation and replication of internal parasites.

Check your learning Can you:

- Evaluate your own needs and preferences, and using the values presented in Table 3 choose the most suitable type of poultry operation.
- Acquire chicks of a type and breed that might be most suitable for your operation.

III HATCHING CHICKS

Purpose

1. Learn how to obtain fertile eggs.
2. Learn how they are incubated (or hatched).

Obtaining Fertile Eggs

Eggs for hatching should come only from hens which have laid above the average of the flock. They should be mated to cockerels that have been the healthiest and fastest growing in their hatch. Both cocks and pullets should be at least 6 months old and in their first year of production for maximum hatchability of the eggs (up to 95%). One active rooster of the lighter breeds can fertilise up to 15 hens; meat type cocks can serve only 10 to 12 hens. The passage of 2 or 3 days is required after putting the cocks and hens together before maximum fertility is achieved; hens continue to lay fertile eggs for as long as 30 days after the males have been removed although fertility declines.

To produce the necessarily clean eggs, hens must lay in nests with clean nesting material such as sawdust, shavings, straw or hay. No more than 3 to 5 hens should use any one nest, otherwise excessive soiling and breaking occurs. Eggs should be gathered 3 to 4 times daily and stored in a cool (10°C to 21°C) but not cold place with 60% to 70% relative humidity. There is a 5% to 10% drop in hatchability in eggs stored longer than one week.

Eggs sold for eating purposes are generally not suitable for hatching.

This is because these eggs usually come from flocks maintained without cockerels, and the eggs may have been coated with a thin layer of light mineral oil. Oil helps preserve the interior egg quality by preventing the escape of moisture and exchange of gases through the shell. Furthermore, the eggs may have deteriorated due to the passage of time, improper storage temperatures or other mishandling.

Natural Incubation

During the 21 day incubation period, brood hens can hatch 8 to 12 chicks from 12 eggs. Commercial hatcheries achieve at least 85% hatch, but, due to a lack of adequate control, home-made incubators yield only 50% to 70% hatch.

Brooding hens should be isolated and in a darkened area provided with a 35 × 35 cm nest that has been cushioned with materials such as wood shavings or straw. They should be regularly supplied with feed and water and not disturbed until after 22 days.

After hatching, the chicks will be cared for by the hen, the chicks eating with the hen. If kept on the ground, they should be provided with a small, well-ventilated coop 0.61 m × 0.91 m and a movable covered wire yard (1.83 m × 3.66 m).

Artificial Incubation

An incubator is an environmental chamber designed to maintain constant temperature, humidity and air quality. It provides a source of heat, moisture and a means of changing air.

Construction   Small, still-air incubators can be constructed in such a way that about 30 eggs on a tray can be suitably incubated. The incubator can be very simple, for example, a cardboard box with a light bulb for heat, a pan of water for moisture, and air holes for ventilation. Or a box can be constructed with no cracks or holes except a small adjustable opening in the front base and one in the back top to provide for some air exchange.

Ventilation   The developing embryos require oxygen for respiration. Ventilation can be increased to reduce temperature and humidity by increasing the size of openings. Some ventilation is always necessary. The oxygen level in the atmosphere (21%) is most desirable; the carbon dioxide concentration should not exceed 0.5%.

Temperature   Heat may be supplied by an electric light bulb or lantern placed above the eggs. If there are hot or cold spots that cannot be corrected, move the eggs to different places on the tray each day to equalise the conditions for all the eggs during the incubation period. Even short exposure to temperatures above 40.5°C is lethal to bird embryos. Opening the incubator increases the ventilation and reduces the temperature and humidity. Although prolonged cooling (beyond 15 minutes) may not kill the embryo, lower temperatures will delay hatching. The temperature should be 38.3°C and should not vary more than 0.28° at any time.

Relative humidity should be 50% to 70% (55% for the first 18 days then 65% to 70% during the last 3 days). In areas of low humidity, a pan of water as large as the egg tray should be placed at the base of the incubator under the eggs. Eggs should not touch the water.

Position the egg by setting the blunt (air cell) end of the egg up at a 45 degree angle from the vertical. Turn the eggs one-half turn an odd number of times (either 3 or 5) each day until the seventeenth day after which they do not need turning. To keep track, mark with a pencil an x on one side and an 0 on the other side of the egg. Jarring is harmful to the egg, and those with cracked shells will not hatch.

Fumigation On the eighteenth day remove the eggs and, while maintaining their temperature, fumigate the incubator. One procedure might be to place 35 g of potassium permanganate in a non-metallic container within the incubator and pour 70 ml of formalin around the crystals. A pungent gas will immediately evolve; close the door and avoid breathing this gas. After about half an hour of fumigation, replace the eggs in the incubator. Fumigation is necessary to reduce disease transmission from the hen via the egg to the chick and from chick to chick while still in the incubator. Pullorum (bacillary white diarrhoea) caused by *Salmonella pullorum* is such a disease and can cause up to 100% mortality in chicks. During fumigation, the eggs can be candled for fertility and embryo mortality.

Candling is done in a darkened room by passing a strong beam of light through the egg and observing the outline and colour of the internal parts. A candler can be made from a box or can in which a 40 to 60 watt light bulb can be placed and a hole about 3 cm in diameter can be cut in one end (Figure 19). Hold each egg before the candler (do not twirl); discard all eggs that are clear or partially clear. Viable embryos will show a clear area only in the air cell, the embryo blocking the passage of light elsewhere. Careful observation will reveal movement of the embryo such as kicking of the legs or jerking motion of the body. A large air cell is necessary to provide air for the chick as it breaks its way through the shell.

Eggs should hatch on day 21. When removed from the incubator the chicks will be soft and wet to the touch. They should be placed in a box in a warm, dry, draught-free room for about 3 hours to 'harden'. They should then be placed in a brooder.

Strict sanitation should be practised throughout. After every hatch all the equipment should be scrubbed with a detergent and disinfected. (See Learning Activity VIII.)

Check your learning   Can you:

• Explain the conditions that must be met in order to get fertile eggs from a breeding flock of chickens. (Management of cocks and hens, gathering and storing the eggs, etc.)
   a. List the conditions that must be met for successful incubation of chicks (temperature, humidity, turning, time, etc.).
   b. Explain how these conditions are met naturally by the brooding hen.
• Construct an incubator capable of incubating 30 eggs.

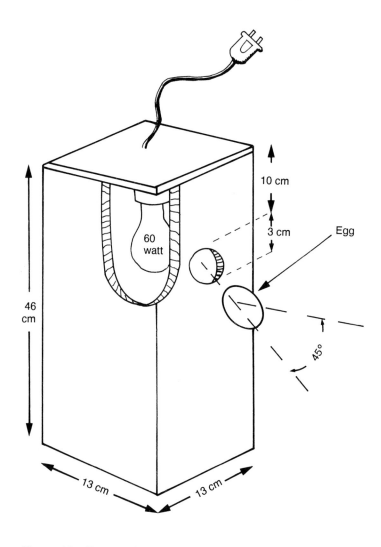

Figure 19   Egg candling device.

118

IV BROODING AND RAISING CHICKS TO MATURITY

Purpose Learn principles of brooding and raising chicks and how these principles can be applied.

1. Obtain chicks from either brood hens or an incubator. If purchased, the chicks should have come from a pullorum-typhoid clean hatchery.
2. Heaters and other equipment and facilities should be checked and put in operating condition prior to receiving the chicks. See Figure 20 for ideas on structures and equipment.
3. Remove all litter and manure. Clean and disinfect brooders and all feeding and watering equipment.
4. A brooder can consist of a large cardboard box or other enclosure located in an area where a moderate temperature (about 18.3°C to 23.9°C) is maintained. A box 0.61 or 0.91 m wide by 0.91 or 1.22 m long will do. A space of 0.74 to 0.84 sq m and 38 to 41 cm deep will house 25 to 35 chicks for 2 to 3 weeks.

   Baby chicks can have too little or too much space. Too little space results in crowding, whereas too much allows chicks to lose the source of heat and feed and water. If a hover is used, each chick should be allowed 45 to 65 sq cm of brooder space. If chicks are started on the floor of a large room make use of a chick guard 30 to 38 cm high to keep them confined in the vicinity of water, feed and warmth.

   The brooder floor can be solid covered with litter or it could be of mesh wire. In the solid system the floor should be covered with 2.5 to 10 cm deep mould-free, absorbent, non-dusty material such as sawdust, wood shavings, rice hulls or straw. This litter should be managed so it will remain dry, being changed as needed. Floor space should provide 697 to 929 sq cm per bird. This system should be used when the temperature is likely to fluctuate.

   Chicks can be started on wire hardware cloth (1.27 cm mesh) that can be nailed onto a 30 × 60 cm to 30 × 90 cm wooden frame. Bottoms of 1.27 cm mesh wire require only 465 to 697 sq cm per bird. Litter or paper under the wire can facilitate cleaning.
5. Heating. An electric heat lamp, electric light bulb, lantern or other heat source securely hung 30 to 50 cm above one end of the box will keep the chicks comfortable and allow them to move in and out of the heat zone. Regulate the brooder temperature to the comfort of the chicks.

   The chick behaviour provides a clue to their comfort. If too cold they complain by chirping and crowding up; if too warm, they will pant and stand with their wings held out from their bodies and will get as far from the heat source as possible. Ideally, the brooder temperature should be 32°C to 35°C for the first week, and 2.8° lower each succeeding week until 21°C is reached or until no artificial heat is required; this takes 3 or 4 weeks. Use a thermometer to measure the proper brooding temperature.

   Care must be exercised in the choice of heat source and its location to prevent fire.

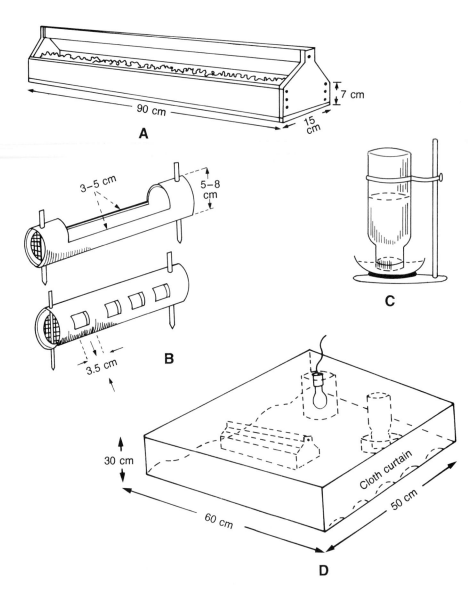

A. Open trough feeder. A bar across the top prevents chicks from getting into the feeder and rotates, not allowing them to rest on it.

B. Bamboo feeders. A section between nodes is opened to provide a strip or a series of holes. The feeders can be secured with pegs or ties.

C. Vacuum type waterer. A bottle that can be secured in an upright position is filled with water and turned upside down in a shallow container.

D. Brooder. A 60 to 100 watt light bulb (or equivalent source of heat) can be inserted in a porcelain socket in a tin. This can be fastened to the centre of the cover, or if no lid is used, suspended from above. A cloth curtain will allow the chicks to go in or out. It is optional whether the waterer and feeder are inside. This brooder can serve 25 to 50 chicks.

Figure 20   Structures and equipment used in brooding and raising chickens.

A hardware cloth or woven bamboo cover on a cardboard box will confine the birds as they grow and learn to fly. Enough ventilation to keep the litter dry without creating draughts should be provided.

6. Waterers can consist of 0.5 or 1 litre fruit jars full of water inverted over a shallow pan. These waterers should be emptied and cleaned daily. Twenty-five chicks should have 1 litre of water daily. At maturity, 25 birds should have 5.7 to 9.5 litres of water daily.

7. Feeders can be shallow box lids or trays for the first day or two until the chicks have learned to eat. A little feed will be wasted before they learn to feed from regular feeders. Chick feeders can be made from bamboo. Using a length of bamboo 5 to 8 cm in diameter between nodes cut out a strip about 1.7 to 2.7 cm wide to provide access for filling and feeding. Or a series of holes 4 cm square can be cut along the length of bamboo. A trough 30 cm in length will provide for 25 chicks for the first week. Feeding troughs should be increased as the birds grow allowing 6.4 to 7.6 cm per bird. At maturity they should have 10 cm of trough space per bird. They should have more feeder space than is needed rather than less. The tops of the feeders and waterers should be level with the back of the birds when standing. They should be so constructed that the birds do not stand or roost on them. Feeders should have a lip, or rolled edge, to prevent feed wastage.

8. There should be enough draught-free ventilation to remove ammonia fumes and keep litter dry. Provide the birds with a minimum of 12 hours of light per day with a minimum light intensity of 10.76 lux (1 foot-candle) at bird height. This is roughly equivalent to a 40 watt light bulb every 3.7 m.

9. When chicks for egg production are 6 to 7 weeks old, or when they have outgrown their brooding quarters, they should be moved to pens where they can remain until adults.

   Separate the sexes as soon as they can be distinguished. Males develop a larger body size and larger, redder combs and wattles than do the pullets. Cockerels (not intended for breeding) should be fed to grow the fastest on the least total amount of feed. They should be ready for slaughter from the seventh to the tenth week, depending on the desired size and type of bird. Table poultry should weigh 1.8 to 2.7 kg in 8 weeks. Laying type cockerels should be dressed at light weights (1.1 to 1.3 kg) since they mature sexually early and the sooner out of the flock, the better. Birds for meat should be raised on solid floors to reduce breast blisters that can occur on wire floors.

10. Birds must be protected from storms and temperature extremes for greatest health and productivity. Protective covering of 0.18 to 0.23 m per bird should be provided. Buildings can be attached to skids and moved about or built permanently with concrete foundations and floors. Pole buildings with dirt floors are sometimes used, but they can have difficult rodent problems and are hard to clean properly.

11. Chicks are most vulnerable to coccidiosis when 2 to 5 weeks of age. During that time the chick starter feed should contain a coccidiostat such as nitrofurazone, or 'Amprol' at the 0.0125% active drug level.

Chicks should be vaccinated for any endemic diseases for which vaccines are available. Follow the instructions that come with the vaccine.

Check your learning   Can you:

● Construct a brooder for 30 chicks.
● Construct feeders and waterers that are necessary for these chicks.

V MAINTAINING A LAYING FLOCK

Purpose   Learn to:

1. Debeak a bird.
2. Cull unproductive birds.
3. Maintain a constant source of eggs.
4. Manage birds in hot weather.

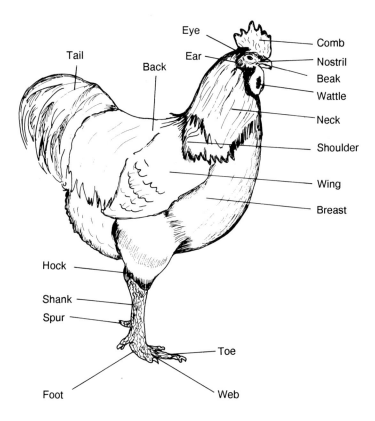

Figure 21   Body parts of a chicken.

122

Controlling Cannibalism

Birds are naturally cannibalistic as seen in toe, feather, vent or tail picking. Any stress such as crowding, bad weather, overheating, irregular feeding, etc., increases the incidence of cannibalism. This situation can be prevented by debeaking which can be done at any age, but preferably at 2 weeks. Debeaking can be accomplished with a pair of scissors, fingernail clippers, sharp knife or with commercial debeaking equipment. Remove half to two-thirds (when measuring from the nostril hole to the tip of the beak) of the upper beak and the tip of the lower beak. If bleeding occurs, it can be stopped by searing the tip of the beak with a red hot iron. Use care not to burn the bird's tongue. If a second debeaking is necessary it should be done at about 12 weeks or when the birds are not in production.

Selecting the Individual Birds

The names applied to the individual parts of a chicken are shown in Figure 21. Culling should be an ongoing process in a chicken flock. From the very beginning, weak, runty, slow growing pullets should be eliminated. Even if unthrifty chicks survive, the chances are poor that they will ever become productive. A hen shows evidence of laying by having:

1. A bright red, smooth, glossy, large comb.
2. A vent that is large, moist, and smooth with a bluish-white colour.
3. Pubic bones that are thin, pliable and spread apart 2 or 3 finger-widths (Figure 22).
4. Pubic bones and keel bone are separated by 3 to 4 finger-widths (Figure 22).
5. Worn, soiled plumage and bleached beak and shank. Good layers show a progressive loss of yellow pigment in beak and legs as laying continues.
6. Late maturing pullets should be culled. Discard pullets not laying by 2 months after the first egg is laid in the flock.

Replacement Programme

A hen is potentially capable of producing 4000 eggs if she could live long enough. She is not in continuous production, however, but after each laying year she will become inactive sexually, lose her feathers (moult) and then grow new ones. With the time lost in the moult and the fact that the number of eggs declines each year of lay, a usual practice is to keep layers only for the first 14 to 16 months of egg production or until moulting begins. The birds are then slaughtered for meat and are replaced.

A constant egg supply requires a second group of layers of a different age. These replacement pullets should be started so that they will have been in production for about a month when the older hens are slaughtered.

If for some reason it is not desirable to purchase or raise replacements so frequently, the productive life of hens can often be extended by their being

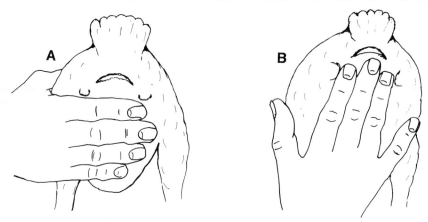

A and B. Laying hen. Pubic and keel bones are flexible and wide apart. This allows for passage of the egg.

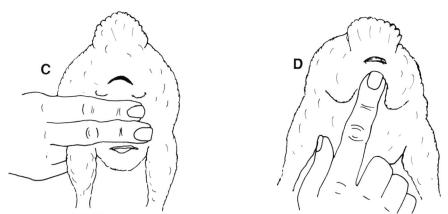

C and D. Non-laying hen. Pubic and keel bones are close together and stiff.

Figure 22   Identifying hens that are laying.

force-moulted after either 8 or 12 months of lay. Cull the flock and moult only the birds that are producing well and are healthy. One of many methods of inducing moult is to turn off all artificial light and remove all drinking water for 2 or 3 days and all feed for 3 to 5 days. Then re-supply water and fully feed a low protein feed such as only maize or milo. After 22 days, birds should resume laying when placed on a balanced feed containing 17% protein. After 8 weeks the birds should reach 50% production and lay for another 8 months.

By moulting at 8-month intervals a flock can be moulted twice for 3 8-month production periods in 24 months. Or, by moulting once after a 12–month lay, 2 production periods, one of 12 and one of 8 months, can be achieved in a total of 20 months. After moulting, 8 months of laying is about all that can be expected. Differences in these management schemes can be illustrated in this way:

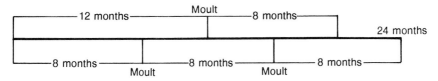

Managing Chickens in Hot Weather

Physiological effects   Chickens, like most other animals, are better adapted to keeping warm than keeping cool. For normal functioning, the body temperature of the bird must be maintained at 41.39°C.

As the environmental temperature rises from the ideal of 12.8°C to 21.1°C the bird tries to maintain body temperature by:

a. reducing the heat input by getting in the shade,
b. reducing the body heat production by reducing its activity and feed intake,
c. increasing the heat loss through evaporation of water.

Birds have no sweat glands to supply moisture for evaporation from the body surface; therefore, they dissipate heat by passing air over the moist surfaces of the respiratory tract (panting). Panting and hyperventilation have an undesirable side effect resulting from the excessive loss of the carbon dioxide that is needed to synthesise calcium carbonate in the uterus; consequently, egg shell quality suffers and soft shelled eggs are often produced.

Failure to maintain normal body temperature (homeostasis) results in a drop in production, heat stress, prostration and death as the temperature rises. High humidity accentuates the effects of atmospheric heat.

Controlling the temperature.

a. Locate buildings in the shade where air movement is unrestricted.
b. Plant grass for at least 6 m around each building.
c. Use open-sided buildings with an east–west orientation so sun will never shine inside. Construct the roof with wide overhangs. Place the roof angle north and south so as not to catch the direct rays of the sun.
d. Provide air movement and evaporative cooling.
e. Provide extra watering devices.
f. Provide extra light early in the morning to induce feeding while the air is cooler. However, once the daylight has been extended in this way, it must be maintained. Shortening the daylight hours by no longer supplying extra light will induce laying hens to moult.
g. In extreme circumstances water can be added directly onto the birds' backs by sprinkling or spraying.

Other Management Factors

1. Put new birds in a clean, dry, well-ventilated shelter, protected from predators, rodents, wild birds and inclement weather.
2. Provide adequate clean, fresh water in fonts that are cleaned at least weekly.
3. Provide fresh, dry, nutritionally balanced feed with adequate space for all birds to eat. Protect this feed from wild birds, rodents, moisture and the bird's own droppings.
4. Keep the poultry house clean and if birds are on the floor maintain 3 to 4 cm of clean, dry litter. Manure, if dry, should be of no concern with mature hens free of intestinal worms.
5. Cull non-laying hens.
6. Prevent feed wastage. Place feeders at proper height and filled not over half-full.
7. Feed according to the age and purpose of the birds. Growing table poultry should be fed differently from laying hens.
8. In the absence of a sloping floor for eggs to roll out of the cage as shown in Figure 1, clean, padded nesting boxes such as those shown in Figure 23 should be supplied.
9. Provide at least 1 nest for 4 to 5 laying hens and make frequent collection of eggs.
10. Light is essential for laying hens. They should have at least 16 hours light per day. Light stimulates egg production. Declining light or shortening days (from 21 June to 21 December) depresses egg production.

Check your learning Can you:

- Debeak a bird to prevent cannibalism.
- By feel and sight, identify a hen that is in production laying eggs. Consider the comb, colour of the shanks and beak, flexibility and spacing of the pubic bones, etc.
- Describe the appearance of a hen in moult. Explain why hens moult.
- What appearance and behavioural changes are shown by birds suffering from heat stress? What procedures should be instituted to reduce the heat load on the bird?

VI CHICKEN NUTRITION

Purpose

1. Learn why chickens need to be fed, their nutritive requirements and some of the factors that influence this need.
2. Learn how these needs can be met; what should be fed and how much.

All animals must consume feedstuffs to provide the nutrients needed for growth, fattening, reproducing and just staying alive (maintenance).

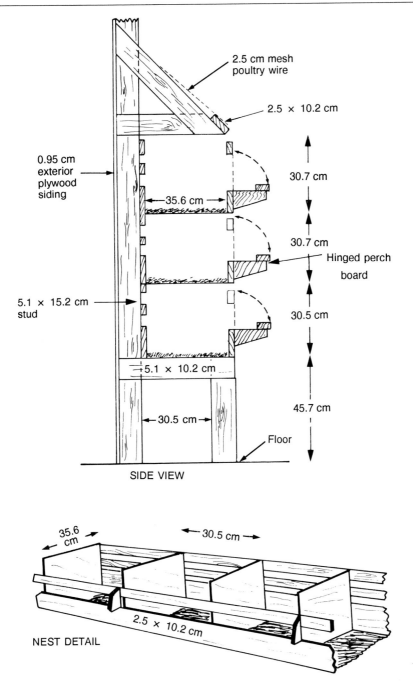

2.5 cm mesh
poultry wire

2.5 × 10.2 cm

0.95 cm
exterior
plywood
siding

30.7 cm

35.6 cm

30.7 cm

Hinged perch
board

5.1 × 15.2 cm
stud

30.5 cm

5.1 × 10.2 cm

30.5 cm

45.7 cm

Floor

SIDE VIEW

35.6 cm

30.5 cm

2.5 × 10.2 cm

NEST DETAIL

Figure 23   Construction detail for a nesting box.

How much of these nutrients are needed will vary with the physiological activities within the bird. Thus, diets differ for baby chicks, growing table poultry, laying pullets or mature hens laying eggs.

A poultry feed should provide the nutrients in proper balance. Incomplete or unbalanced rations not only reduce performance but may also cause nutritional diseases. Nutritive requirements are rather specific because of the nature of the digestive tract and amount of metabolic work performed. A newly hatched chick can double its weight in 5 days and reach one-half of its mature weight in 2 months. It takes man 150 days to double his weight and 115 to 145 months to reach one half of his mature weight. Young birds double their weight several times during the early weeks of their lives if their nutritional and other needs are met.

In formulating a ration consideration should be given to the following:

1. The protein needed to provide the essential amino acids is measured in grams. Methionine and lysine are the essential amino acids that are most likely to be the limiting factors.
2. Energy is derived from carbohydrates, fats, and, when in surplus, protein. Energy is measured for poultry in terms of kilocalories of metabolisable energy (kcal ME). Fat has 2.25 times more energy than carbohydrate or protein. Each gram of carbohydrate and protein supplies 4 kcal ME and each gram of fat provides 9 kcal.
3. The vitamins of special concern are niacin, riboflavin, vitamin B_{12} and vitamin A.
4. Minerals, especially calcium, phosphorus and salt.
5. Water must be supplied at all times.
6. The ration must be palatable so the birds will eat it in sufficient amounts to meet production needs. The bird becomes profitable only after it consumes more feed than is needed just to stay alive. The diet should allow the bird to perform at its genetic potential.
7. Using a variety of ingredients in the ration reduces the risk of any deficiencies. Also it allows the limited use of some materials such as cottonseed meal, that at higher levels would be harmful. (Cottonseed meal should not be more than 5% of the total ration.)
8. Chickens eat to meet their energy needs. It is thus possible to regulate the intake of all nutrients (except water) by regulating the concentration of the nutrients in relation to the energy in the feed. The birds can consume enough nutrients to be most productive if their ration provides at least 2750 to 3000 kcal ME for every kg of feed consumed. The ration is then balanced so that when the amount of feed consumed meets this energy need, it will at the same time meet other nutrient needs as well.

The following data illustrate that the amount of feed eaten by a hen varies with the energy concentrations in the feed and the environmental temperature. In each case the bird is consuming 315 kcal ME in the winter and 280 kcal ME in the summer. Temperatures below or above the zone of thermal neutrality (14.4°C to 25.6°C) will require extra energy in order for the bird to maintain its normal body temperature of 41.1°C to 41.7°C.

Feed Energy Level kcal/kg	Daily Feed Intake Winter g/hen	Summer g/hen
2640	119	106
2860	110	96
3080	101	91

9. Per unit of body weight, smaller birds have a higher maintenance requirement. However, because of their greater size larger birds require more total feed for body maintenance than do smaller birds even when they are producing the same number of eggs. This principle is demonstrated in Table 7. Birds in 70% production being fed a ration containing 2900 kcal ME per kg will require 75 g if the hen weighs 1.0 kg and 132 g if she weighs 3.0 kg.

Table 7   Feed Requirement of Hens as Influenced by Body Weight and Level of Egg Production (2900 kcal ME/kg of feed)

Body Weight kg	Feed (g/Hen/Day) at 4 Production Levels			
	0%	50%	70%	90%
1.0	45	66	75	83
1.5	61	82	91	100
2.0	75	97	105	114
2.5	89	111	119	128
3.0	102	123	132	141

Data based on Nutrient Requirements of Poultry. NRC, 1984.

Smaller and younger birds require less feed to add an increment of weight. This is because the weight that is added by larger and older birds contains a higher proportion of fat, and fat requires more energy to deposit than does protein. Regarding body composition changes, for each unit increase in body fat there is a concurrent approximate decrease of 0.73 units in body water and 0.03 units in body protein. Within breeds, up to about three-quarters of mature weight, the composition of the body is a function of the body weight, not age. During the growth period, the composition of any gain in animals of the same body weight will be the same whether this weight was put on slowly or quickly.

10. Growth requires extra feed. The protein content of the feed should be higher for young chicks and for meat type birds (Table 5). Also, since males grow more rapidly than females, males can be grown to a given size with less time and total feed than is required for pullets. Furthermore, pullets just coming into production are still growing and will require more feed than mature hens. Generally, birds raised for meat

are fed higher energy rations to get maximum growth in the least time. The growth rate and the amount of feed required is estimated in Table 8. Animals should be managed to grow rapidly in order to reduce the overhead expense of maintenance.

Table 8   Growth Rate and Feed Consumption of Layer Chickens and Table Poultry

Age, weeks	Body Weight, g		Cumulative Feed Needed, g	
	Male	Female	Male	Female
Layer type[1]				
4	270	270	725	630
8	690	620	2600	2550
10	965	790	3750	3320
12	1240	950	4900	4180
16	1450	1160	6200	6020
20	1700	1360	8600	7860
24	1800	1500	11000	10100
30	—	1725	—	12970
50	—	1870	—	16040
70	—	1900	—	19030
Table poultry type[2]				
2	320	300	360	350
4	860	790	1305	1205
6	1690	1430	3025	2650
8	2520	2060	5330	4530
10	3330	2640	8000	6580

Values based on Nutrient Requirements of Poultry. NRC, 1984

[1] Based on feed containing 2900 kcal ME/kg.
[2] Based on feed containing 3200 kcal ME/kg.

About 9.0 kg of feed are required to rear a laying pullet to laying age in 22 weeks. The average kg feed required to add a kg of growth for the different bird types up to 10 weeks of age are:

Layer males	3.9
Layer females	4.2
Table poultry males	2.4
Table poultry females	2.5

Some of the nutrient needs for chicks are summarised in Table 9. To reflect the changing needs, the concentration of both protein and energy in the feed declines as the birds mature.

Starter and grower feeds should contain a coccidiostat to control coccidiosis in the growing chicks. Amprol at the 0.0125% active drug level might be useful. The salt should be fortified with the trace minerals and a vitamin supplement should contain vitamin A, riboflavin, niacin and vitamin B_{12}, which are the vitamins most likely to be lacking in sufficient amounts.

Feeders should never be allowed to become empty. Add fresh feed twice daily, but add only slightly more than the birds will eat before the next feeding. To reduce waste do not fill the feeders full.

Table 9   Feed Programme for Chicks

Age	Protein %	ME kcal/kg	Ca %	P %
Egg Type Chicks:				
0 to 5 weeks	20	2800–3100	1.0	0.6
6 to 14 weeks	17	2800–3100	1.0	0.6
15 to 20 weeks	15	2600–3100	1.0	0.6
Meat Type Chicks:				
0 to 5 weeks	24	3200–3400	1.0	0.6
6 to 9 weeks	20	3200–3400	1.0	0.6

11. Egg production. From the following data note that at zero production each bird consumes 72 g of feed or 72×2950 kcal ME/kg = 213 kcal metabolisable energy, which is the maintenance requirement of a 1.81 kg chicken. Each egg laid requires about 120 kcal ME above maintenance (a large egg itself contains about 90 kcal and 6.8 g of protein). Thus, each day an egg is laid the hen requires 213 + 120 = 333 kcal. Hens in a flock at 80% production would each need 213 + (0.80×120) = 309 kcal ME (378 kcal DE) daily.

The feed must, then, not only support maintenance and activity but also must replace the nutrients which were lost in producing the egg. Although increased egg production requires more feed per hen, the amount of feed required per egg produced (efficiency) is increased. To illustrate, assume that 1.81 kg hens are fed feed containing 3000 kcal ME/kg:

Production Rate %	Daily Feed Intake g/hen	Feed Efficiency kg feed/doz eggs
0	72	—
20	80	4.76
40	88	2.63
60	96	1.91
80	104	1.54

Feed accounts for the largest single item of cost in the production of eggs, constituting 50% to 70% of the total costs. Normally, a mature laying type hen will consume 36.3 kg of feed per year and produce 20 dozen eggs; thus, more than 1.81 kg of feed is required to produce a dozen eggs. An example of a balanced ration for hens in 75% to 80% lay is:

Ingredient

53.44 kg Yellow maize
10.0 kg Milo (sorghum)
19.5 kg Soybean meal (47.5% protein)
 7.0 kg Meat and bone meal (50% protein)
 1.0 kg Stabilised grease (optional)
 8.0 kg Limestone or oyster shells
 0.6 kg Dicalcium phosphate (optional)
 0.15 kg Salt (trace mineralised)
 0.06 kg dl-Methionine
 0.25 kg Layer vitamin premix

100   kg

Calculated Analysis

2900 kcal/kg Metabolisable energy
18.0% Crude protein
 4.0% Crude fat
 2.5% Crude fibre
 3.89% Calcium
 0.50% Available phosphorus
 0.97% Lysine
 0.68% Methionine and cystine

Commercially formulated and mixed feeds will usually be similar to the above in composition. Although probably costing more per kg than home grown feeds, commercial feeds with their variety of ingredients and expertise in formulation may sometimes be the more profitable to feed especially to hens with a high production potential.

If the feed has been ground, grit is not a necessary ingredient. However, grit becomes essential if whole grains are fed.

In the absence of commercial feeds, chickens can be sustained (but with a lower production) if fed all the cracked grain they will eat along with enough protein supplement to provide 18% protein plus 0.5% salt, a free choice calcium source and lucerne, weeds or some other acceptable green feed. These plant materials can be tied into a bundle and hung in the cage each day so that it can be eaten by the birds. Table scraps might also be used, but table scraps and grain alone are not satisfactory chicken feed.

A simpler mix based on either wheat or maize for energy and whole soybeans for protein has been demonstrated to support production at an acceptable level. The soybeans must be roasted or otherwise treated to inactivate the trypsin inhibitor.

Simple Wheat and Maize Based Rations for Chickens

Ingredient	Layer Chicken Diet 16% Protein	Table Poultry Chicken Diet 20% Protein
Wheat	66.9	60.1
Soybeans (roasted)	22.4	34.9
CaHCO₃	7.5	1.8
Dicalcium P	2.0	2.0
Vitamin and mineral premix	0.7	0.7
Salt	0.5	0.5
Total	100.0	100.0

Ingredient	Layer Chicken Diet 16% Protein	Table Poultry Chicken Diet 20% Protein
Maize	61.8	55.5
Soybeans (roasted)	27.5	39.5
CaHCO₃	7.5	1.8
Dicalcium P	2.0	2.0
Vitamin and mineral premix	0.7	0.7
Salt	0.5	0.5
Total	100.0	100.0

Check your learning   Can you:

● List the classes of nutrients needed by the chicken.
● Identify the factors that influence the total amount of feed needed by one bird. What factors influence the proportions of nutrients in the feed?
● a. Determine the total quantity of feed needed for one year for:
   a. Growing 12 pullets to laying.
   b. Maintaining 24 laying hens in 80% production.
   c. Growing 6 sets of 22 table poultry chicks to 1.8 kg in 10 weeks.

VII FEEDING CHICKENS

Purpose   From available feedstuffs, learn to formulate a ration that will meet the nutritive needs of chickens.

Balancing poultry rations is largely a matter of correcting the nutritional deficiencies of the feedstuffs used as energy sources (grains, fats, cassava, etc.). Considering the needs of laying type birds, for example, a starter ration containing 20% to 22% protein is fed for the first 8 weeks, then a

19% protein growing ration is fed, followed at 18 weeks by a layer ration of 15% to 17% protein. Caloric density of these rations should be 2600 to 3100 kcal ME/kg. Table poultry should be encouraged to grow faster with a 23% protein diet for the first 5 weeks and 20% protein to 9 weeks and a caloric density of 3100 to 3450 kcal ME/kg in the feed. Salt (preferably trace mineralised) at 0.5%, calcium at 1.0% and available phosphorus at 0.6% should be in the growing rations. Laying rations should contain an additional 2% calcium (3% in total). The calcium supplement could be oyster shell, calcium carbonate, bone meal, tri-calcium phosphate or some other calcium source being fed free choice.

The Pearson square can be used to establish proportions of 2 feeds (or two mixtures of feeds) to mix providing that 1 of the 2 ingredients is above and the other below the desired level. In optimal rations protein is derived from many sources; however, to illustrate the principle, and at the same time note compositional changes as various ingredients are added, assume that we initially wanted to mix corn at 9% protein with soybean meal at 40% protein to get a feed containing 20% protein.

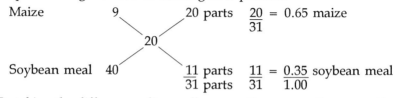

Maize   9   20 parts $\frac{20}{31}$ = 0.65 maize

                         20

Soybean meal   40 $\frac{11 \text{ parts}}{31 \text{ parts}}$ $\frac{11}{31}$ $\frac{0.35}{1.00}$ soybean meal

By taking the differences diagonally across the square, the proportions of each ingredient are obtained. Thus, 20 parts (65%) of maize and 11 parts (35%) of soybean meal will yield a mixture of 20% protein.

The metabolisable energy concentration of this mixture is:

65% maize   @ 3350 kcal = 2177 kcal/kg
35% soybean meal   @ 2230 kcal = 780 kcal/kg

                     Total 2957 kcal/kg

This feed with 2957 kcal metabolisable energy per kg, when supplemented with about 5% of non-energy yielding minerals and vitamins, would yield 95% × 2957 or 2809 kcal/kg which is suitable for growing chicks.

Admixing a fibrous feed such as lucerne meal (1370 kcal/kg) would further decrease the nutrient concentration. Assuming 5% as the maximum suitable level of lucerne and if lucerne contained 17% protein, the protein content of the ration would remain essentially unchanged. The energy level would be reduced, however.

95% maize–soybean meal mix @ 2809 kcal = 2669
5% lucerne meal   @ 1370 kcal =   68

                            2737 kcal ME/kg

If this feed were to be fed to layers, additional calcium to raise the level to at least 2.75% would be needed. The organic ingredients of the feed already supply:

58.5 kg maize @ 0.03% Ca   0.02 kg Ca
31.5 kg soybean meal @ 0.29% Ca   0.09
 5.0 kg lucerne meal @ 1.33% Ca   0.07
(5.0 kg premix)   —
_____
100.0 kg   0.18 kg Ca

Using limestone @ 33.7% calcium:

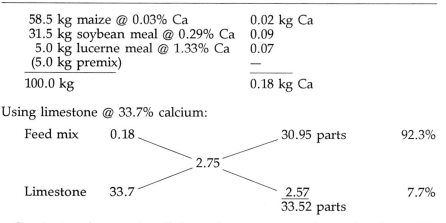

Feed mix   0.18   30.95 parts   92.3%

                    2.75

Limestone   33.7   2.57   7.7%
                       33.52 parts

Continuing the exercise, if these changes were to be made, about 18% of the original feed mixture (65% maize and 35% soybean meal) would have been replaced. Then, 100 kg of the end product would supply:

kg/100kg	Ingredient	Protein %	Protein kg	Energy kcal ME		Calcium %	Calcium kg
53.5	Maize	(9)*	4.80	(3350)	1776	(0.03)	0.02
28.8	Soybean meal	(40)	11.48	(2230)	640	(0.29)	0.08
5.0	Lucerne meal	(17)	0.85	(1370)	68	(1.33)	0.07
5.0	Premix	—	—	—	—	—	—
7.7	Limestone	—	—	—	—	(33.7)	2.60
100.0			17.13		2484		2.77

* Composition values for feedstuffs are shown in brackets. Rounding can introduce slight discrepancies.

Whereas the desired level of calcium is achieved there is a concurrent drop in protein and energy content of the feed. These levels could have been maintained by adding additional high protein and high energy feedstuffs.

It can be seen that this method of feed formulation is not only somewhat tedious but imprecise when trying to balance more than one nutrient factor. Similar results can be obtained through the use of simultaneous equations. Computer programmes have been written that can provide precise information for feed mixtures. However, even without the use of computers using simultaneous equations, while recognising the relative, indefinite nature of feed analysis tables, the above procedure serves quite well in practical feeding situations.

Recall that comparable feedstuffs can be substituted in a ration; that is, wheat, barley or sorghum can substitute for maize, groundnut meal for soybean meal, and regular cottonseed meal can make up to 5% of the ration at the expense of soybean meal, etc. Within limits, excess protein, minerals and vitamins can be tolerated by the birds, although such excesses might increase the cost of the feed.

Check your learning   Can you:

- Formulate a chick starter mash that will meet the stated nutrient requirements.
- Formulate a grower ration for table poultry.
- Formulate a layer ration.
- Demonstrate that these rations supply the requirements of protein, energy and calcium for all the above chickens.

VIII SANITATION AND HYGIENE FOR CHICKENS

Purpose   To understand the need for sanitation and disease prevention among animals.

Diseases cause illness and death. Healthy animals are free from disease.

Look through a microscope and see some of the tiny forms of life that are present but cannot be seen with the unaided eye. Most of these do not hurt us; in fact, many are beneficial. Some of these organisms, however, make us sick (or cause diseases). Those forms of life that are harmful are called pathogens. They are of several different forms such as viruses, bacteria, moulds and protozoa. Diseases might even be caused by worms and insects.

There are many diseases among animals. Only a few examples are fowl typhoid, fowl cholera, Newcastle and coccidiosis. Some diseases like tuberculosis (in cattle but not in chickens), brucellosis and anthrax can be transmitted from animals to man.

Pathogenic organisms are so small that they cannot be seen without the aid of a microscope; they are called microorganisms. In order to cause damage, the pathogens must first enter the body through the digestive tract, respiratory system, intact or injured skin, or other body openings. Pathogens are transferred from sick to other animals in manure, body secretions and coughing.

To prevent sick animals from making the healthy ones ill also, they should be kept separated (quarantined). When new animals are brought onto the farm, they should be kept separate for 2 to 3 weeks so that if they should get sick they will not contaminate all others. Poultry should not be exchanged among farms. When birds are taken off the farm they should not be returned.

Wild birds and animals frequently carry disease organisms which can cause the disease in domestic animals. Therefore, wild birds and animals should be kept away not only from the farm animals but from their feed and pens as well.

If the domestic animals are well fed, protected and healthy, they are more resistant to the pathogens. The natural resistance of animals to some diseases can be increased to higher levels by vaccination.

In addition to invisible pathogens, internal and external parasites can also cause damage to animals, and so must be controlled. These parasites include protozoa, roundworms, tapeworms, flukes, mites, lice, ticks and fleas. The purpose of sanitation is to destroy or weaken the pathogens and parasites before they can harm the animal. Chemicals that destroy pathogens are called disinfectants or sanitising agents. Anthelmintics and insecticides are used to control worms and insects.

Inadequate sanitation is one of the most important causes of health problems in poultry production. Sanitation begins with cleanliness. Manure, rotting feed, feathers and rubbish should not be allowed to collect, but should be either tilled into the land for fertiliser or burned.

Pens and cages in which sick animals have been kept should be thoroughly cleaned and disinfected before other animals are placed in them. Also cages, nesting boxes, etc., should be cleaned and disinfected after each use or season. Scrape away caked dirt and manure and wash with a brush and water before rinsing with a disinfectant.

Disinfectants can be prepared by several methods and should be applied to clean surfaces. Manure and other dirt can neutralise the effects of disinfectants, because the disinfectant reacts with the dirt rather than the pathogen; therefore, the equipment should be cleaned before the disinfectant is applied. Disinfectants are more effective when applied hot. Some effective disinfectants are:

30 g or ml sodium hypochlorite (laundry bleach) dissolved in 1 litre water.
1 g lye (NaOH) in 45 ml water.
1 ml cresol per 32 ml water (a detergent will facilitate emulsification).
Chlorhexidine preparations, to be used as directed.
Idophore preparations, directions on the package should be followed.

When animals are kept confined on the same land for long periods of time or when large numbers of animals are concentrated on small areas, disease organisms and internal and external parasites build up in numbers and virulence. The pattern can be broken by moving animals to fresh uncontaminated ground or buildings and by tilling the ground.

Check your learning   Can you:

- Name the different types of organisms (pathogens) that cause disease (example: bacteria).
- Some pathogenic microorganisms and internal and external parasites cause diseases. What are some of the similarities and differences in the way they affect the animal?
- State the manner in which a disease can be communicated from one animal to another.
- Explain how quarantine can help control disease.
- Explain why wild birds and animals should be kept away from domestic animals and their feed.

- Explain why equipment and surfaces must be cleaned before applying a disinfectant.
- Identify at least one disinfectant and explain how it can be used in farm sanitation.
- Describe how to break the buildup cycle of pathogens and of internal and external parasites.

IX CHICKEN FLOCK HEALTH

Purpose

1. Learn the nature of diseases.
2. Learn how to prevent disease (sanitation).
3. Learn some diseases that are commonly encountered.

Careful sanitation, adequate space, comfortable housing, proper feeding and generally good management are the best preventatives to keep the flock free of diseases and pests. In addition, certain diseases that are endemic to an area require vaccination to control. The poultryman should come to recognise the appearance and behaviour (symptoms) of the birds when they are diseased. General symptoms of disease are coughing, sneezing, difficulty in breathing, watery eyes, droopiness, bloody/abnormal appearing faeces, a sudden drop in feed and water consumption and a drop in egg laying. The owner of a small flock might treat for lice and mites and a few other diseases. Individual birds when dead, or those that appear diseased should be killed promptly, necropsied and then either burned or buried deeply enough so that they cannot be unearthed by predators or pets. Often, should a highly virulent disease appear in the flock, the best solution might be to eliminate all the birds for several months and completely clean and disinfect the equipment, buildings and area before new birds are introduced.

Review the material presented earlier under 'Health and Disease'. If available, check with a veterinary surgeon or other qualified person to learn which conditions in the locality might present health problems and how they might be controlled.

Sacrifice a bird thought to be infected with parasites, perform a post mortem examination and identify any abnormalities.

Check your learning   Can you:

- Name the principal diseases in the area and state their symptoms and means of control.
- Examine under magnification faecal droppings from birds thought to be infested with parasites. Try to identify oocites or worms. Explain how to recognise the presence of any internal parasites and state measures to be taken in their control.
- Examine a bird for external parasites. Explain how to recognise the presence of any external parasites and state measures to be taken in their control.

X POULTRY EGGS AND MEAT

Purpose

1. Learn how to care for eggs.
2. Learn to slaughter chickens and properly handle the meat.

Eggs

An egg should have a strong, regular and clean shell, thick, clear and firm albumen (egg white), and yolk that is evenly coloured, well centred and free from blood and meat spots. These qualities do not materially affect the nutritive value of the egg, but do affect aesthetic appeal. The structural parts of an egg are shown in Figure 24A. The size of air cell increases as the egg ages; fresh eggs will have smaller air cells (Figure 24B).

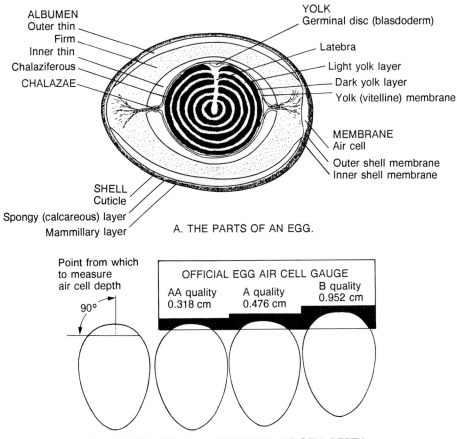

ALBUMEN
Outer thin
Firm
Inner thin
Chalaziferous
CHALAZAE

YOLK
Germinal disc (blasdoderm)
Latebra
Light yolk layer
Dark yolk layer
Yolk (vitelline) membrane

MEMBRANE
Air cell
Outer shell membrane
Inner shell membrane

SHELL
Cuticle
Spongy (calcareous) layer
Mammillary layer

A. THE PARTS OF AN EGG.

Point from which
to measure
air cell depth

90°

OFFICIAL EGG AIR CELL GAUGE

| AA quality 0.318 cm | A quality 0.476 cm | B quality 0.952 cm |

B. EGG QUALITY AS MEASURED BY AIR CELL DEPTH.

Figure 24   The structure of the egg in cross-section and how egg quality is measured by air cell depth.

Egg quality depends on genetics as well as feeding and management of the bird and care of the eggs. Heredity influences shell colour and thickness, egg size, quantity of the thick white (albumen), quantity of eggs produced and egg spots.

The breaking strength of the egg shell is affected not only by the hen's breeding, but by her feed, especially vitamin D and calcium, as well as her age, freedom from disease and by hot weather. Shelless eggs can be laid by hens exposed to heat stress, as well as calcium and/or vitamin D deficiency.

Yellow grains and green feeds produce a yolk that is dark yellow to orange coloured. Green grass and gossypol from cottonseed meal can cause yolks to be reddish to olive colour.

The high percentage of thick white (albumen) of quality eggs can be reduced by genetics, sickness, and older age of the layer, as well as by improper care of the egg after it is laid.

The following rules should be followed to ensure top quality market eggs:

1. Provide adequate non-staining, dry nesting material with no more than 4 birds per nest for birds not in cages.
2. Gather eggs frequently, 3 to 4 times daily, and clean any dirty ones immediately. This is because the foreign material adhering to the shell contains bacteria that can penetrate the shell and contaminate the contents of the egg. Those eggs only slightly dirty can be cleaned by buffing with a dry abrasive such as emery cloth. The dirtier eggs will require wet washing. The following should be observed in this process:
   a. Wash with water warmer than the egg, otherwise water and soil will be drawn through the shell pores into the inside as the egg cools.
   b. Use a detergent-sanitiser that will (1) loosen the soil, (2) kill bacteria, (3) have no foreign odours to impart to the egg.
   c. With a clean cloth dipped in this solution remove the dirt from the egg.
   d. Rinse the egg with warm water.
   e. Dry the egg.
3. Apply a light coat of thin, odourless, colourless and tasteless mineral oil.
4. Cool eggs immediately after gathering to 12.8°C and store at 75% to 80% relative humidity. (Eggs for hatching can be stored 5°C warmer.)
5. Candle eggs either before or after cooling to cull out those with checked or cracked shells and blood and meat spots. Defective eggs should not be stored.

Candling is a method by which the inside condition of an intact egg can be determined. This is done by passing a strong light through the egg. The candler consists of a light-proof, 5-sided box (open bottom) with 3 cm openings on one side and on the top. The light source can consist of a 40 to 60 watt electric light bulb or similar light source (Figure 19, p. 118). In a darkened room, permit the light to pass through the egg by holding it against the hole, viewing it at a 45 degree angle, thus looking at the egg and not the light.

a. Consider the shell first. Look for small cracks (checks) in the shell, and observe the shell texture, shape and cleanliness.
b. Look at the blunt end and examine the air cell. The air cell is formed soon after the egg is laid. Water within the egg is lost to the atmosphere and is replaced by air. The two shell membranes part to fill the void that is thus created. The air cell increases in size with the passage of time after the egg is laid. Fresher eggs have smaller air cells.
c. Turn the egg while it is against the candler opening. The yolk should be seen only as a slight shadow with limited freedom of motion. This indicates a firm albumen holding the yolk in place. The sharper the yolk shadow, the thinner the albumen, indicating a poorer quality of egg. Any red or black discolouration in the albumen can be caused by some blood or bits of tissue (blood or meat spots). These defects are aesthetic in nature and do not reduce the nutritive value of the egg.
6. Eggs should be stored with the small ends down, keeping the air cell in the large end on top. Avoid storing eggs together with sources of pungent odours such as petroleum, onions and garlic because eggs absorb such odours.

Meat

1. Birds should be starved for 3 to 4 hours before slaughter to reduce the contents of the crop and digestive tract.
2. Kill the bird by severing the vertebra and blood vessels in the neck while keeping the skin intact. Hold the chicken upside down by the legs with the left hand, stretch the bird across the front of the body by grasping the head with the other hand, with the base of the neck in the palm between the thumb and forefinger and the beak between the third and fourth finger. By sharply stretching the chicken, the head is forced backward and the cervical vertebra and vessels severed. Suspend the chicken by its feet head downward for about 5 minutes to allow the blood to accumulate in the neck. Considerable force may be required to break the necks of older birds, in which case decapitation with an axe might be more appropriate. Or with a little skill and a sharp, pointed killing knife, with the chicken confined, the top of the head can be held in one hand and the knife pushed into the soft tissue of the throat from side to side at the base of the skull; the cutting edge facing away from the skull. The bird will bleed freely if the jugular vein and its shunt have been cut (Figure 25).

Poor bleeding is directly responsible for poor keeping quality, the development of undesirable flavours and an unappetising appearance of the carcass.

After hanging remove the chicken, pull or cut off the head and the carcass is now ready for removal of the feathers and viscera.
3. Feathers can be loosened for plucking by holding the bird by the feet and sloshing it up and down for 1 to 1.5 minutes in hot water (54.5°C

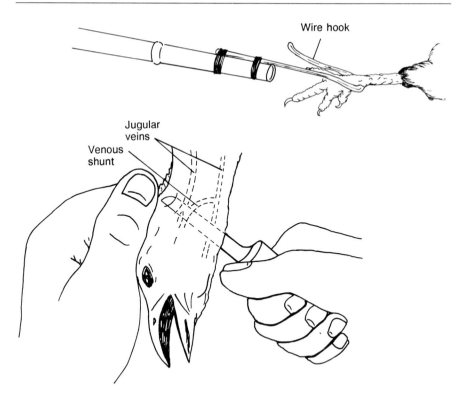

Figure 25   Chicken restraint, killing and bleeding.

for fryers and 71.1°C for older birds). When the main tail and flight
feathers can be pulled easily, remove the bird from the water and pull
the feathers as quickly as possible, beginning with the large flight and
tail feathers.

4. With a knife remove the oil sac (preening gland) on the back at the base
of the tail.

5. Split the neck skin, and remove the oesophagus, crop and windpipe
by pulling them away from the neck skin. Cut the oesophagus where it
enters the body cavity just to the rear of the crop; tie off the oesophagus
to prevent the contents from contaminating the neck and breast. Pull
out the windpipe where it enters the body cavity.

6. With the bird on its back, cut around the vent and gently pull out 10
to 15 cm of the intestine. Open the abdomen and remove the viscera
with the hand. Exercise care that the intestines and gall bladder are
not punctured and contaminate the meat. Separate the liver, gizzard
and heart from the viscera. Cut away the gall bladder, keeping it below
the liver. Split the thick muscle of the gizzard lengthwise down to the
inner lining. Pull out the lining, trying to leave the gizzard contents

142

intact. Remove the lungs and ovaries or testes that are located in the bird's back.

7. Chill the carcass and giblets to below 4.5°C as quickly as possible. Maintain this temperature until the chicken is cooked. The meat will stay fresh for 3 to 4 days. Unlike beef or pork, chicken meat deteriorates very rapidly at room temperature. Mature chickens are too tough to grill, roast or fry, so must be boiled.

Check your learning   Can you:

- List steps to follow in order to produce quality eggs.
- Construct a candler and candle 6 eggs.
- Kill, pluck, eviscerate and cut a chicken into serviceable parts.

9
RABBIT PRODUCTION PRACTICES

Special consideration should be given to the production practices of rabbits because of their fast growth and their near total dependence on human care. Kits double their weight in 6 days and a doe can produce 10 times her body weight in weaned kits within a year. Those who care for rabbits (especially women and children) are the ones toward whom any instruction and training should be directed. The reproduction, growth and feed conversion values cited herein are generally those noted in better managed rabbitries in the temperate zone. It is improbable that these same results are to be achieved in the tropical areas of the world because of disease conditions, poor breeding stock, insufficient feed, adverse climate and lack of management skills. It is important that this difference be recognised in order to avoid disappointments arising from unrealistic expectations. A less intense subsistence-scale operation is often the most beneficial and profitable system under village conditions.

SELECTION AND BREEDING

There are many rabbit breeds, varying externally as to body size, shape and length, and the length, density, colour and colour distribution of the hair. Fancy breeds have been developed only to meet the conformation specifications of predetermined breed standards. Some breeds, however, have been developed primarily for production traits such as rate and efficiency of gain and total meat or wool produced. Breeds native to the developing countries are typically small and slow growing, which improves their chances of survival in a harsh environment but reduces their productivity. The principal body regions of a rabbit are shown in Figure 26.

Choosing a Type and Breed
1. Select a type and breed that has done well in the area.
2. If meat is desired, choose a meat breed of medium size weighing 3.5 to 5 kg at maturity. The New Zealand (white, red or black) or the Californian (white with coloured ears, nose and feet) are generally suitable. Within these breeds, smaller strains (2.5 to 3 kg) with larger ears should be selected in warmer climates. These might be crossed with native breeds to develop desirable crossbreeds. Full grown males are generally 10% to 15% lighter than females; in most other species the male is larger.

3. The short haired (20 mm) Rex rabbit was developed from a mutation in Europe in the 1920s. Its non-shedding, attractive pelt can be made into fur jackets and coats. Originally, the Rex was a small animal weighing only 1.8 kg. It now has been out-crossed with larger breeds to produce animals weighing 2.7 to 3.6 kg. This makes the Rex a dual purpose meat and fur animal. Short hair is thought to make regulation of body heat easier in hot weather.

4. Long haired Angora rabbits produce wool for weaving clothing, or their pelts are used for fur production. Their hair length is about 200 mm. Only such long hair is suitable for shearing and weaving. Higher hair density and thicker skins increase the value of the pelts for fur processing. However, these characteristics can reduce the dressed carcass weight and thus the secondary value of the animal for meat.

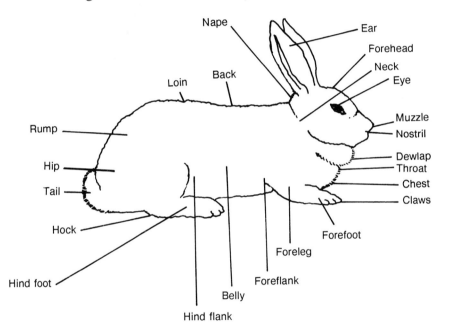

Figure 26   Body parts of a rabbit.

CHOOSING THE INDIVIDUAL ANIMAL

Conformation

While there is little scientific evidence proving that one body proportion and shape is more efficient in production than another, there are aspects of body form that are generally considered desirable:

Head   Broad, symmetrical (not pear shaped) with evenly carried ears. Relatively large ears help heat dissipation in hot climates.

145

Teeth and jaws   A genetic defect keeps the incisors from meeting by a shortening of either the upper or lower jaw (malocclusion). The middle incisors grow continuously and if they are not kept worn down they can grow so long that they prevent feeding. Since malocclusion is basically due to a recessive gene, its appearance in an animal would indicate that both parents (although they themselves appeared normal) carried the undesirable gene and, therefore, should be culled from the breeding herd. Malocclusion can result from other than genetic defects.

Back   Wide and long providing a large back and loin for meat. The loin should be slightly higher than the hips. Swayback (low over the mid-section) should be avoided. When viewed from the rear, the top should be well rounded, and when viewed from the top, there should be a slight taper from front to back.

Hips   Wide, rounded and long, standing slightly above the back level. A smooth, filled-out, rounded rump is desirable.

Legs   Straight, neither cow-hocked nor pigeon-toed. Legs provide the principal source of meat; it is thought that there should be about 4 finger widths between the hind legs of does to provide adequate birthing space.

Hind feet   Large, capable of supporting the body weight. Sore hocks tend to develop more often on small feet or on feet with poor fur density on the foot pads.

Belly   Filled in but not pot-bellied. There should be at least 8 evenly developed and spaced teats.

Body length   Long and smooth, meat types should be 38 to 43 cm in length.

Body shape   A stout stocky body shape decreases the proportion of more valuable parts of the carcass (back and legs) in favour of the front parts. A wide, deep body is desired.

Generally, smaller animals have a higher performance and survival value in a tropical climate.

Other desirable traits

In choosing the individual animal consider also a rapid growth rate, prolificacy and mothering ability. As with all classes of livestock, foundation individuals should be selected from disease-free rabbitries and from animals that have performed well under conditions of feeding and management similar to those to which they will be exposed. Foundation bucks should be unrelated to the does to reduce the harmful effects of inbreeding, viz., a decreased size and vigour and the appearance of harmful recessive characteristics.

Does

Does should have at least 8 evenly placed and developed teats. The number of teats is genetically determined. Does with less than 8 teats generally produce less milk and have difficulty supporting more offspring than they have teats; therefore, they should not be used in breeding. Does should be selected on the basis of their ability to produce heavy litters and to breed back after kindling. At the end of the year those does producing a total weight of fryers below the herd average are candidates for culling. More specifically, a doe should be selected on the basis of the following traits:

1. She is easily bred and conceives at the first service to a fertile buck.
2. She makes a good nest and maintains this nest throughout the first 10 days after kindling.
3. She kindles 6 to 10 kits in the kindling box, on average. Does that fail to kindle in the box twice should be culled.
4. She is a good milk producer as indicated by the total 21-day and weaning weights of the litter.
5. Her offspring grow rapidly as indicated by their 8-week individual weights. There is a high correlation between daily weight gain and efficiency of feed conversion.
6. The number of her offspring reaching market age annually is above the average for the herd.
7. She rebreeds readily and raises litters in excess of the herd average.
8. She maintains her health and thriftiness, recovering easily after kindling and nursing a litter.

Bucks

Bucks should possess high libido (sex drive) and high fertility, as indicated by a high proportion of fertile matings. The average litter size born should exceed 7; the litters, in particular, should have a high rate and efficiency of growth to market weight, be disease resistant, and have high fur and carcass quality. Young bucks for breeding should be chosen from dams that are superior for maternal characters and be from sires that excel in the market weights of their offspring at a standard age (56 days). Bucks as well as does should be free of genetic disease (e.g. hydrocephalus, buck teeth or malocclusion). They should have thick foot pads with sound hocks, and have dense fur with good texture. Desirable breeding animals will then be more capable of transmitting these characteristics to their offspring. In herds consisting of one or more breeding does 2 bucks should be available for breeding in case one proves to be sterile. Replacement bucks should be kept from the fastest growing and largest in the litter.

BREEDING PLANS

The small number of animals associated with small-scale agriculture drastically limits what can be accomplished in an individual breeding programme. A few things should be considered, however. The production of larger, faster growing litters is a management goal. Since the heritability of reproduction traits (number of young born and reared, interval between litters, etc.) is relatively low, best results could be achieved by breeding crossbred does to purebred bucks. A buck of a different breed is used to mate each new generation of does in a meat producing programme.

Traits that can be improved through selection and breeding (have a higher heritability estimate) include: body size, daily gain in weight, feed efficiency, number of teats, ability to lactate, amount of milk produced, the quality and quantity of wool, and pelt colour and quality.

In native breeding situations there has been a tendency of counter-selecting for size. By slaughtering the largest and heaviest of the litter, the smaller, less developed individuals have been left to reproduce the herd. The problem is often compounded by inbreeding which results in both smaller size and decreased vigour in the animals. The ill effects of inbreeding can be reduced by using bucks from unrelated herds. It should be borne in mind that introducing unrelated animals from other farms carries the danger of introducing disease as well. For this reason, animals newly brought to the farm should be quarantined for 30 days and only if then found free of disease should they be introduced to the other animals. Exchanged breeding animals should be young because such animals are more adaptable to changes in climate, feed and management when 3 to 5 months old.

Where outcrossing is not feasible and sufficient animals and interest are present, other breeding options are available to the commercial breeder. One mating scheme that could help fix some desirable traits in the herd would be to select from the litter the largest, fastest growing and best formed male and female; then, mate the buck back to his dam and the doe to her sire. If there is yet another pair that shows exceptional size and vigour, these could be mated together. Such mating patterns increase the homozygosity of both the good and bad aspects. For this reason, to avoid inbreeding problems, concurrent strict culling of animals with bad points needs to be practised. By breeding parent to offspring it is possible to determine which line has the greater influence in producing the desirable (and undesirable) traits. Use the line with the most good points without serious defects for future breeding. By following this pattern for a few generations it would be possible to 'fix' the desired type in the herd.

REPRODUCTION

Mating

The genital organs of the buck are essentially the same as in other mammals (Figure 27). The cylindrical penis is situated on the ventral (belly) side of the anus and when erect is 3 to 5 cm long. Since there is a positive correlation between testicle size and spermatogenesis, testicle size can be a useful factor in selecting breeding bucks. Shortly before puberty, at 3 to 4 months of age, the testicles descend into the scrotum. The buck can retract both testes into the abdominal cavity if excited. The testicles can be forced back into the scrotum if the abdomen is pressed by the palm of

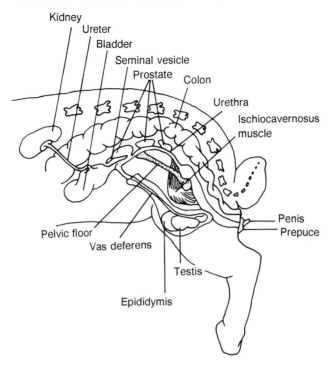

Figure 27   Reproductive organs of a male rabbit.

the hand. Males that retain the testicles in the abdominal cavity beyond 12 weeks should be culled as this is an inherited trait. Sexing prior to that time can be accomplished by noting the greater distance between the anus and penis (males) than exists between the anus and vulva (females). Genitals of infant males appear as a round 'O', whereas the female vulva appears as an elongated opening (Figure 28).

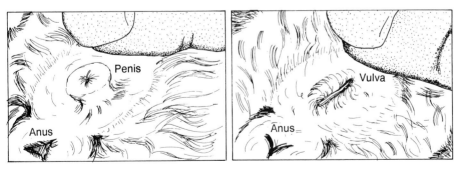

Figure 28   Comparative external structures of the male and female rabbit.

149

To determine the sex of a mature rabbit, place the animal on its back holding it firmly around the shoulders with one hand and support the hips with the palm of the investigating hand. Hold the tail between the first and second finger and extend the thumb up and apply a slight pressure in front of the genital organ located between the thumb and the anus. This pressure should expose the organ as a slit-like vulvular opening in the female or a slightly protruding and rounded penis in the male.

At first mating, medium sized does should be at least 5 months and bucks 6 months of age; or they should weigh 80% of their mature weight. Body weight is better than age as a criterion for deciding when animals are ready for breeding. The vulva should have a pinkish red colour which becomes darker with maturity. Smaller breeds reach puberty earlier than do the larger ones. Both males and females should be in good condition, neither too fat nor too thin. Does should be caged individually during the breeding period to reduce riding and pseudopregnancy. To prevent her from attacking the buck because of her protective instinct, the doe should be taken to the buck's hutch for mating. Her receptivity is demonstrated by a hind leg jerking action if the vulva is rubbed. There is also a vulvular swelling with a bright purple/red colour, as well as a willingness to mate. If the doe is receptive, the buck will mount immediately. The erect penis is carried forward along the abdomen. After the buck makes a few thrusting motions, the doe raises her hindquarters, allowing the insertion of the penis. The sperm in the ejaculate is followed by a gelatinous plug from the accessory glands. If mating is completed, the buck might give a faint cry and fall either backwards or on his side. Bucks in good condition should be able satisfactorily to serve 10 to 15 does at the rate of 2 to 5 does per week; many bucks can be used daily for several months without losing libido and fertility. Adult males can serve more does than younger bucks can.

Some breeders, but not all, feel that increased conception rate and litter size can be achieved by allowing 2 services at the time of mating; this maximally stimulates the doe to ovulate and makes use of the second ejaculation of the buck which usually contains more sperm. Other producers report that increased litter size results from remating the doe 8 to 12 hours after the first mating. In any case, if the doe does not accept the buck, she should be removed and returned to the buck after 2 to 4 days. Breeding 2 or more does at the same time permits later fostering of kits to equalise number of kits per lactating doe.

The doe has 2 uteri, each having a distinct horn. They do not join to form a common body of the uterus but each has its own cervix opening into a common vagina (Figure 29). Five to 10 ova-containing follicles usually develop on each ovary and remain active producing oestrogen for 12 to 14 days. After this time, if ovulation has not occurred, these follicles will degenerate with a corresponding drop in oestrogen resulting in a refusal to mate. After about 4 days a new wave of follicles will begin to produce oestrogen and the doe will again become receptive to mating. Thus, the doe has a 16 to 18 day cycle with about 12 to 14 days of receptivity and 4 days in which she refuses to mate. She shows her unwillingness by thumping

150

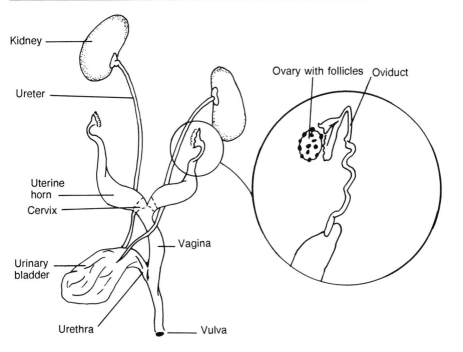

Figure 29 Reproductive organs of a female rabbit.

her back feet or other signs of aggressiveness. Such periods of rejection are extremely variable due to individual differences, sexual stimulation and environmental factors such as nutrition, light and temperature. A new set of ova will be present immediately after kindling and on the seventeenth to twentieth day. If she is in good condition and there is a high level of management, the doe can be rebred at any of these times.

Ovulation in the doe occurs about 10 hours after being subjected to some external stimulus such as sexual copulation or other intense excitement. Ovulation can result from being mounted by other does or by being stroked and petted. If insemination does not accompany this ovulation, pseudopregnancy may result. In pseudopregnancy the doe behaves as if she were pregnant. Usually by day 17 of pseudopregnancy the doe can be successfully rebred.

A doe can be force mated by restraining her with one hand grasping the ears and shoulder skin, and placing the other hand beneath her body with the thumb and forefinger pushing backward and upward on either side of the vulva. This pushes her tail over her back, making it easier for the buck to mount. Experienced bucks will generally make a successful mating, although reduced conception rates or fewer kits result from such forced matings.

Techniques for artificial insemination have been developed. There are many similarities to other classes of livestock. Successful insemination requires 0.5 to 1 ml of diluted semen containing 1 million live or 2 million total sperm. Ten to 50 inseminations is the maximum that can currently be made from one ejaculate. No successful techniques for extended storage of semen have been developed. Careful attention to all details is necessary for success.

Pregnancy Check

Only about 70% conception rate occurs with the initial mating. To save time in the breeding cycle, it is as well to detect those does who fail to conceive as early as possible, so they can be re-bred. Ten to 12 days after breeding, does should be tested for pregnancy by palpating the developing foeti in the uteri, a method referred to as abdominal palpation.

Back the doe into a corner of her own pen, or place her on a table that provides a firm footing. Calmly restrain her by the ears and skin over her shoulders with one hand and place the other hand under the body just in front of the pelvis. As gestation continues the foeti will be lower and more forward in the abdominal cavity. Place the forefinger and thumb on opposite sides of the abdominal cavity. Raise the hind quarters until just the tips of her rear feet are touching the floor. Wait for her to relax her abdominal muscles. Gently feel for the embryos which will have a diameter of 12 mm at 9 days and 22 mm at 13 days of pregnancy. At 14 days the marble-sized foeti are easy to feel and the likelihood of damage from palpating minimal. The foetus begins to elongate at 15 to 18 days of pregnancy. Palpation for pregnancy should not be attempted after the eighteenth day because of possible damage to the placental attachment to the uterine wall. If no foeti are present, rebreed the doe. Faecal pellets will not be mistaken for embryos if it is remembered that the uterus lies on the bottom of the abdominal cavity whereas the large intestine is above, nearer the backbone (Figure 30).

Kindling

The rabbit has an average 31 to 32 day gestation period, normally ranging from 29 to 35 days. Larger litters with correspondingly smaller kits are carried for shorter periods than are litters with fewer kits that are larger (weighing up to 100 grams each), the weight of average kits being 50 to 60 g.

Twenty-seven or 28 days after breeding, the doe should be provided with a disinfected kindling box. The purpose of the box is to provide protection for the kits. However, the doe is apparently unaware of ever stepping on her kits, and in poorly designed boxes she may cause injury when jumping into the box.

Nest box management is important. Different types of kindling boxes are illustrated in Figure 31. Traditional boxes have approximate dimensions of 30 cm × 45 cm and 40 cm high with an opening 15 cm above the floor. The proper size should be about 5 cm longer and wider than the doe when in a sitting position. If the nest box is too large, the doe may sit in it and foul it with faeces and urine. Does seem to prefer a box with at least a partial

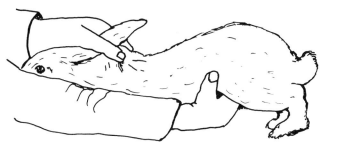

Figure 30 Determining pregnancy by palpation.

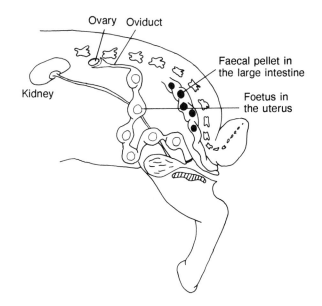

top cover and one that they can jump down into. However, for best results the nest box should be without a top and situated below the floor of the cage, the box being 15 cm deep. This 'drop' or 'subterranean' box allows the operator to see in to remove dead kits etc. An important advantage is that if the kits do crawl out, they fall back into the box as they move around the cage. Does do not return their young to the nest, and if kits become separated from each other, the doe will nurse either those inside the box or those outside, but usually not both. Bedding from such materials as chopped straw or soft shavings should be clean, dry, pliable, short and absorbent. It should be mixable with the hair that the doe will pull from her own body.

153

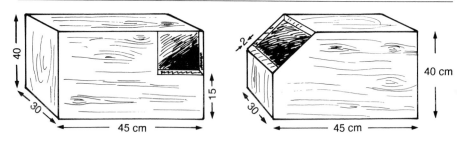

Figure 31   Kindling boxes.

If the doe does not pull enough hair herself to keep the kits warm, additional hair can be pulled from her body and added to the box. Often, fur can be saved from other does and used when needed. Hair that is between 20 and 70 mm in length is most suitable for nest building. Too much hair in the nest is detrimental in hot weather. Hair that is too long can become entangled around appendages of kits and cause strangulation. A flat depression in the middle of the nesting box floor will help keep the litter together. The doe licks the kits to stimulate defaecation and urination; then she eats the faeces. The urine falls into the bedding, suggesting the necessity for dry absorbent material or a means of drainage in the box floor.

At kindling time, especially, does should not be disturbed by unusual noises or the presence of dogs, cats, exuberant children, strangers, etc. If the doe is young (first litter), or if disturbed, she may drop her kits over the cage floor and/or eat some of them. If this behaviour is observed at

154

the second kindling, the doe should be culled. Timely nest building, a good quality nest, milk yield and other aspects of maternal care generally improve up to the third or fourth litter that the doe drops.

Does usually kindle at night. Kits are born hairless, blind and deaf. Generally, there are no complications at parturition even with breech delivery. With the larger kits of small litters, kindling may be delayed a day or two. Kindling should be completed after 30 minutes with kits being delivered at 1 to 5 minute intervals following birth contractions of 10 to 20 seconds. By sitting in a bent-over position the doe directs her vaginal opening forward and slightly sideways. She can then assist the birthing of the young and tear open any closed amniotic sacs and bite through the umbilical cord as the kits are born. Sometimes the doe may injure or amputate an appendage which might cause the doe to eat the entire young kit. After birth, each kit is licked by the doe to clean and stimulate it. If kept warm and dry, kits do not require nursing for the first 48 hours. Usually, however, the kit may nurse immediately but, after this, it will have opportunity to nurse only once or twice daily. The doe normally has 8 teats. The milking ability of the doe can be assessed by the total weight of the litter at 21 days.

The naked, blind and deaf newborn kits are confined to their nest for the first 2 or 3 weeks. Losses during this time are quite high (up to 30% or more). Reasons for the losses include stillbirth, low birth-weight, delayed birth, infectious diseases such as salmonellosis and listeriosis, cannibalism; even singleton births are at risk. Many deaths of young kits occur due to ruptured bladders from lack of stimulation by the doe.

The litter size varies from 1 to 24 with an average of 6 to 10. With longer birth intervals the litter size increases slightly (up to 10%). There is less chance for more than 8 (or the number of teats on the doe) in the litter surviving because the doe nurses only once or twice daily and the competition among the kits for available teats is great; the larger more vigorous kits survive at the expense of the weaker ones. This justifies the use of the number of teats as a criteria for selecting breeding animals. Young rabbits die if they miss nursing for 1 or at most 3 consecutive days. If it is not possible to put excess kits in with another doe whose kits are within 1 or 2 days of the same size (fostering), those in excess of the number of teats on the doe might be killed. Size or sex should be the basis of selection. Some does with exceptional milk producing capacity can easily raise more kits than they have teats, although at weaning individual kits will be smaller.

After a doe's entrance into the nest, young kits require only a few seconds to attach to a nipple. They use odour cues from milk in the nipple area and some tactile cues to find their place to nurse. For this reason it is difficult to hand raise neonates; however, orphaned kits might be salvaged with consistent care. Rabbit milk is very rich, providing the nutrients needed for rapid growth by the kits. Rabbit milk has about 30% total solids containing 11% to 13% protein, 12% to 13% fat, 2.5% mineral and 2% sugar. It can be roughly duplicated by beating a hen's egg into 250 ml of 4% cow or goat milk. This is fed very slowly using a gastric tube, syringe or doll's nursing bottle. Up to 5 ml is given the first day, divided

155

into 2 to 3 feedings. This amount is gradually increased to 15 ml per day by the second week and 25 ml by the third week in 2 daily feedings. The average birth-weight of a kit for medium and heavy breeds is 50 to 60 g. The birth-weight of kits is reduced by a large number in the litter, by being the doe's first litter or by malnutrition in the last third of gestation. After kindling, the doe generally eats the placenta. Later, after the doe has left the nest box, the litter should be quietly inspected with a minimum of handling and any deformed or dead kits should be removed to prevent their decomposing bodies from becoming sources of infection. Some does will abandon or kill their young if they have been handled during the first few days, or if the does are unduly disturbed. A properly fed doe can be expected to produce daily 30 g of milk per kg of body weight. Thus a 4 kg doe will produce 100 to 200 g (average 120 g) of milk. Milk yield reaches its maximum between 18 and 23 days postpartum. Secretion declines by two-thirds after 6 weeks and ceases after 8 weeks. The kits will have full stomachs if they are successfully nursing. For the first 3 weeks of life 1.7 to 2 g of milk is necessary to produce 1 g of growth in the kits. Since does will accept kits not their own, especially if within 1 to 2 days age difference, underfed kits can be shifted to does having more milk available. Any introduction should be made a few hours before the nursing time. Also it would be well in fostering to mask odours by touching the doe's nose and the young with perfume or other such material.

POST KINDLING CARE

Lactation by the doe is a very physically demanding activity. Five to 7 days after kindling, corresponding to the pattern of increased milk production, additional feed should be made available.

The desire to reduce the pressure of the milk in the mammary gland seems to be the primary motivation for the doe to nurse. The doe spends only 3 to 5 minutes once or twice each day (usually in the morning) nursing her young. This is not much time for the young to find a teat and nurse up to about one-third of their body weight. They are blind for the first 9 to 10 days during which time pheromones guide them to the teat. Weaker animals are at a disadvantage in finding the most productive teats. A properly designed nesting box with short fibre bedding is helpful in locating feed. Some producers have reduced mortality in their litters by locking the doe out of the nesting box except for a 15 to 30 minute period each morning. The nest box should be removed after the kits are 2 and before 3 weeks of age. This reduces the incidence of disease.

Young rabbits start eating solid feed earlier if the doe produces less milk. Generally, kits depend on milk entirely for the first 2 weeks; at 3 weeks they get 17% of their energy from solid feed, at 4 weeks 45%, at 5 weeks 60%, at 6 weeks 73%, and at 8 weeks they are weaned to obtain 100% of their nutrients from dry feed.

However, if a balanced feed of 2600 kcal DE/kg and 16% protein is available, by 28 days the young should be consuming enough solid feed

156

so that they can be weaned. Kits can be weaned at 6 weeks of age if fed only good quality roughage. They could be weaned earlier if weighing over 350 g and a grain supplement is available. One to 2 days before weaning the doe's feed should be drastically reduced. This reduces the milk secretion of the doe, lessening the likelihood of mastitis. She should be fed at about 50% of the normal ration for the 3 days past weaning. A reduced milk supply will also encourage the young to consume more dry feed. Weaning is thought by some to be less traumatic if the doe is removed and the young remain together in the cage for 3 or 4 days before being transferred to new surroundings. Then, 3 to 4 months after weaning, when the young does are 5 months old, they should be ready for breeding.

If disturbed by predators or otherwise, does generally respond by stamping their back feet, the sound of which should signal a need to investigate. At the same time, if the doe is in the kindling box, this stamping can do damage to young kits.

In tropical areas, under reasonable conditions of feeding and management, a doe should produce 3 to 5 litters totalling at least 20 offspring each year. Usually the does are fed only roughage and are bred 6 to 8 weeks after kindling and the kits suckle the doe for 6 to 8 weeks. By improving the feeding, breeding and other aspects of management, large, healthy does can operate at a more intense level. If does and weanlings are fed concentrates, the does are re-bred 18 days after kindling and the kits are large enough to be weaned by 6 weeks of age. The young does will have reached 80% of their mature weight and be bred by 5 months of age. This could allow for 7 to 8 litters per year. This system is summarised in Figure 32A and part A of Figure 32B.

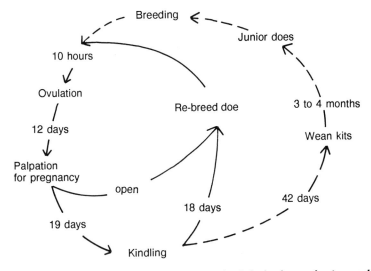

Figure 32A *The breeding cycle. This time schedule is dependent on adequate feed, water and other care as well as animals with the genetic potential to produce under this vigorous management scheme.*

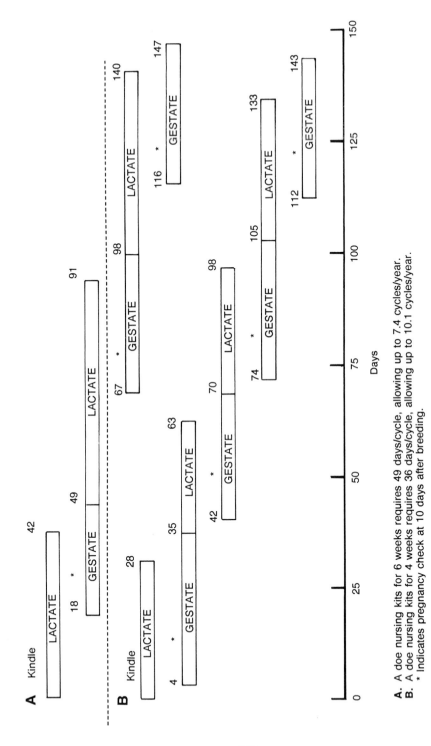

A. A doe nursing kits for 6 weeks requires 49 days/cycle, allowing up to 7.4 cycles/year.
B. A doe nursing kits for 4 weeks requires 36 days/cycle, allowing up to 10.1 cycles/year.
* Indicates pregnancy check at 10 days after breeding.

*Figure 32B Scheduling the reproductive activities of does lactating for 6 or 4 weeks.
In both cases a 7-day rest interval is provided between lactation periods.*

In the most intensive production situation (shown in Figure 32B) as in a commercial rabbitry with vigorous animals in temperate zones, does could be re-bred on the same day they kindle or the first 2 to 4 days postpartum. It is at this time that the highest conception rate is obtained. Weaning, then, should be at 25 days postpartum. There should be at least 5 days recuperation between weaning and the next kindling. Since lactation has priority over gestation for the use of nutrients, simultaneous lactation and pregnancy causes a reduced birth-weight and an increased stillbirth rate. Breeding on day 2 will provide a breeding interval of 33 days allowing a theoretical maximum of 11 litters per year. Such intensive breeding programmes are not universally recommended unless conditions are optimal since such programmes can be counter-productive. Schedules for management planning must recognise that not every doe will conceive at every mating; a conception rate of 4 in 5 matings is usual for natural mating.

Rabbits grown as fryers are generally fed and managed to reach optimum size within the least possible time, using the least amount of feed and experiencing the fewest deaths. Sometimes taking longer to reach the desired weight will increase the total number of animals reaching slaughter size. Eight to 12 weeks are generally required, with the final weight depending largely on the breed and level of feeding.

FEEDING

A sketch of the digestive system of a rabbit is shown in Figure 11. The rabbit has 28 teeth with the incisors continually growing 0.5 to 2 cm monthly. If incisors do not meet properly (as in malocclusion) they are not worn down, and their uncontrolled growth will ultimately prevent the animal from eating.

Rabbits consume most types of hays, greens and grains which can be produced on the farm. Grains should be rolled or coarsely ground making them more palatable with a minimum of dust. Finely ground particles in the gut decrease intestinal motility resulting in constipation and impaction. Grains can be successfully replaced by some of the by-products of the milling industry such as wheat mill-run and rice bran. Roughages and concentrates should be fed concurrently for best results. Feed may be consumed in a mash form, but rabbits prefer their feed to be pelleted where this process is available. Pelleting not only increases intake but prevents sorting as well. This quality can be either an advantage or disadvantage, depending on the quality of the ingredients. The rabbit is highly selective in its eating, having definite preferences. Thus, palatability is important to maximise intake and animal performance and to minimise wastage of feed scratched out of the feeder. If given the opportunity rabbits eat only the more nutritional and easily digested shoot tips and leaves of plants, leaving the stalks and stems; if forced to eat the entire plant, feed intake and animal production suffer. Rabbits prefer legumes to grasses.

Coprophagy is habitually practised by rabbits (Figure 33). After they begin to eat solid feed at 3 to 4 weeks of age, rabbits produce 2 kinds of faecal

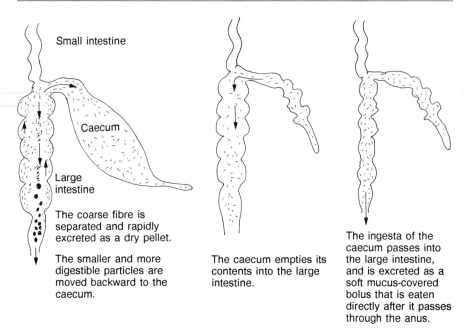

Figure 33   Coprophagy in the rabbit.

matter. One is a dry fibrous pellet and the other is soft mucous-covered faeces. The digestive strategy of the rabbit is to separate and quickly void the coarser, more fibrous portions of the feed. The finer particles and less-fibrous portions of the feed are collected in the caecum for fermentation. This separation in the large intestine is made possible by the presence of out-pouchings or sacculations together with sets of longitudinal muscles (haustra). Normal peristalsis moves the ingesta forward. At the same time the longitudinal muscles in the sacculations contract moving the adjacent material backward toward the caecum. The lighter, coarser, fibrous particles remain in the centre of the gut and are moved onward to the rectum. The heavier, smaller, non-fibrous particles and fluids move to the outside to be shifted backwards to the caecum. After the fibrous faecal pellets have been voided from the colon, the caecum contracts, voiding its contents which pass along the colon to be reingested directly as they come from the anus. (This process usually occurs early in the morning.) Thus, this non-fibrous fraction is recirculated through the digestive system allowing for greater enzymatic and fermentive digestibility. This system of separating and eliminating the fibre while re-circulating the more digestible fractions permits the rabbit to thrive on roughage without the necessity of a large fermentation vat. The situation is further improved by the ability of the rabbit to select and eat only the more digestible parts of the plant. Coprophagy also permits the digestion of some microbial cell bodies that are associated with fermentation in the caecum and large intestine. These

160

cells thus provide some of the essential amino acids and most of the B and K vitamins that are needed by the rabbit.

Volatile fatty acids (VFA) from fermentation provide 10% to 12% of the daily caloric requirement. Multiple passage of ingesta through the digestive tract can partially account for the high digestibility of forage protein (75% for lucerne protein). On the other hand, crude fibre is very poorly digested; its digestibility is only about 16% in rabbits compared to 44%, 41%, 33% and 22% for goats, horses, guinea pigs and pigs, respectively.

Even though rabbits do not digest fibre well, they do have a fibre (fibrosity) requirement. Not only must the chemical entity of crude fibre be present, but some of the particles must be at least 1 cm in length. Generally there is an inverse relationship between the amount of fibre and the concentration of energy in the feed; as the fibre content of a feed decreases the energy concentration goes up. The rabbit has the ability to vary its intake greatly in order to meet its energy requirement; consequently it can easily adjust to fluctuating energy concentrations in the feed. However, a level of 14% to 16% crude fibre and 2500 kcal of digestible energy (DE) per kg of feed is a standard recommendation. A level of 20% to 22% fibre in a maintenance diet can be reduced to 12% to 15% in a growing, gestating and lactating ration.

Problems with enteritis, ketosis, obesity, slow growth and poor reproductive performance are frequently associated with rations that have less than 14% crude fibre or feeds that are too finely ground or feeds with more than 3000 kcal DE/kg. On the other hand, in fattening rabbits, a crude fibre level of over 17% can limit the energy intake and reduce the growth rate.

It should be remembered: **rabbits do not perform well on high energy diets!** Some people have tried to feed rabbits exclusively on commercial pig or poultry rations; but this is a grossly improper feeding practice that will produce disastrous results. However, such feedstuffs can successfully be fed to rabbits if fed as a supplement to a primarily roughage ration.

Health problems aside, up to 10% of dried faeces can be added to the rabbit diet without affecting the feed conversion. Although rabbits will consume grass hays, they have a preference for legumes, especially lucerne. In fact, rabbits can thrive on good quality lucerne hay as a sole feed. Succulent feeds such as freshly cut legumes, grasses, weeds and forbs are relished by rabbits. Frequently these fresh feeds have a tonic effect. In fact, rabbits have been known to eat over 50% of their liveweight daily in fresh, succulent feed. However, if wild plants are fed they must be relatively free from toxins. Uneaten residues in the feed box should be removed daily; rabbits should not be expected to eat spoiled feed. Preference for cereal grains is in the decreasing order of oats, barley and wheat with maize being the least liked. Oats are a high fibre, medium protein cereal grain that is less likely to cause digestive upsets. Oats serve as a useful feed when enteritis is present.

In addition to forages, rabbits can be fed crop residues and kitchen wastes. From the farm might come surplus or damaged bananas, plantains, avocados, palm nuts, mangos, etc. Table scraps include peelings, husks, hulls and plate scrapings that would contain fats and oils and salt. Such supplementation can make feeding of grain unnecessary providing it meets the energy and nutrient needs of the animal. Rapid and drastic changes in the composition of the feed should be avoided because of the limited capacity of the rabbit to adapt and of its susceptibility to enteritis.

For meat animals adequate energy and nutrient intake is especially important before puberty. With medium weight breeds it is possible to get daily gains of 35 to 45 g by feeding a balanced pelleted feed, but a gain of only 15 to 30 g is possible if nothing but mediocre roughage is fed. Maximum growth depends on a diet providing at least 2500 kcal DE/kg because young animals cannot eat enough of the bulkier feeds to meet energy needs for maximum production. Generally, from a performance point of view, there is no advantage in feeding a diet more concentrated than 2500 kcal DE/kg. However, economically, energy can often be produced more cheaply in the form of grain. Yet, at the same time, there is an upper limit on the amount of grain that can be fed. Due to the peculiarities of the herbivorous digestive system, rations more concentrated than 3000 kcal DE/kg might produce digestive and physiological upsets. The quantity of feed required to put on a unit of body weight varies widely with the energy concentration in the feed. Obviously, to produce a unit of gain, it will require more of a feed containing 2200 kcal DE/kg than one containing 2800 kcal. Thus it requires 2.5 to 4 kg of feed to add a kg of body weight. About 7 kg of feed is required to produce a rabbit with a slaughter weight of 2 kg.

A typical response of weanling rabbits under optimal conditions might be the following:

Dietary DE	Average daily gain	Average daily feed intake	Feed/gain	Daily DE intake
3000 kcal/kg	38.1 g	117 g	3.1	350 kcal
or 2500 kcal/kg	38.1 g	140 g	3.67	350 kcal

Rabbits should be fed regularly once daily, preferably in the evening. To avoid obesity it may be necessary to limit the energy intake of bucks and open, non-lactating does to a level equivalent to maintenance. Bucks in service should be fed at a level of 115% maintenance.

For the reproducing doe there are 4 periods of time when adequate nutrition is especially important:

1. One week before mating to increase the number of ova shed for fertilisation (flushing). Some managers question the effectiveness of this practice with rabbits;
2. One week after breeding to reduce the chance of embryonic death;
3. One week before kindling to support about 90% of the foetal growth;
4. Three weeks beyond kindling so the doe can produce the most milk. During the first 2 weeks of this period the kits are totally dependent on this milk.

Growing rabbits as well as the nursing does should have all the balanced feed they will eat. Rabbits consume as many as 40 meals per day, 60% to 70% of which are taken after dark and before daylight. Feeding can take from 2 to 6 hours depending on the accessibility and concentration of the feed. The higher figure would apply if sparse roughage only were available. Feed intake declines with increasing environmental temperatures; thus animals tend to take advantage of the cooler nocturnal hours to eat.

In an exceptional commercial rabbitry, since nursing will reduce the dry feed intake of the young, if a balanced, pelleted diet of at least 2500 kcal DE/kg is available, kits could be weaned at 3 weeks. However, if only roughage will be available, as is frequently the case in developing countries, kits should not be weaned until at least 6 weeks of age. After 8 weeks of age the young rabbits develop the capability of adjusting their intake to meet their caloric needs. Diets containing as low as 2100 kcal DE/kg can then be fed. However, the growth rate will be slowed when feeding at these low energy levels. Limiting the intake of feed or feeding an unbalanced diet may cause rabbits to eat their own hair, resulting in the formation of hair balls in the stomach.

Rabbit rations should contain up to 10% but not over 20% fat. Excessive fat in the ration leads to rancidity and a reduction in the time that the feed can be stored.

Some calculations and assumptions have been made in determining the energy requirements of rabbits. Daily energy requirements for maintenance have been calculated using the formula:

$$\text{kcal DE} = (\text{kg body wt})^{0.75} \times 70 \times 1/0.82 \times 1.25$$

The standard formula for basal metabolism is based on 70 times the metabolic body weight and is expressed in terms of metabolisable energy. Digestible energy is obtained by dividing ME by 0.82. This value is increased by a factor of 25% to account for energy used in activity at maintenance and in thermoregulation. This value would be increased yet further if animals were more active or exposed to cold or hot weather. 9.5 kcal DE is required for each gram of growth and fattening.

The maintenance ration should be doubled during the last 10 days of pregnancy to provide for the additional nutrients needed for foetal growth.

Requirements for lactation will depend on the quantity and composition of the milk produced. Assuming that there is twice the energy in rabbit milk as there is in goat milk, 2.92 kcal DE is required to produce each gram. A 4 kg doe producing 200 g of milk daily will require 584 kcal DE for production and 301 kcal DE for maintenance for a total of 885 kcal DE daily.

Minimum crude protein values vary from 12% for maintenance to 17.5% for lactating does. Crude protein of 16% to 18% in the diet is acceptable in most cases; however, it is reported that levels of up to 22% have produced better reproductive performance, larger litters and lower mortality. Caecal fermentation and coprophagy satisfy some but not all of the essential amino acid requirements of the mature rabbit; the addition of lysine and methionine to roughage rations fed to mature rabbits increased their performance. All essential amino acids (especially lysine and methionine required as 0.65% and 0.6% of the diet, respectively) need to be included in the diet of the very young. Angora rabbits have a higher requirement of methionine and other sulphur-bearing amino acids to meet the increased need for wool production. Rabbits are sensitive to an amino acid imbalance. Slight excesses of threonine, lysine, isoleucine and phenylalanine depress growth. A deficiency of dietary protein results in a reduced feed intake. An energy deficiency, however, has the opposite effect: it increases consumption. Excess protein can serve as an energy source, although usually it is more expensive.

There should be a balanced relationship established between protein and energy. A feed should contain 0.06 to 0.07 g of digestible protein per kcal DE. A ration containing 2500 kcal DE/kg should contain 15% to 18% crude protein. Anything that would reduce the intake such as temperatures above 25°C (75°F) would require an increase in the protein proportion in that feed to ensure an adequate intake of protein. Protein and energy are likely to be deficient in an all-grass-roughage diet.

Although there may be a theoretical advantage to the feeding of urea, or other non-protein nitrogen sources to supplement a low protein diet, there is no conclusive evidence that this is helpful. Similarly there is no advantage to supplemental methionine on a 16% protein diet of lucerne and wheat mill-run.

Vitamin supplementation in rabbit diets is generally not necessary, especially if some green feeds are included. These feeds will provide the vitamins A and E (the vitamins most likely to be deficient). The vitamin D requirement of the rabbit is very low. Some of the B vitamins and vitamin K are synthesised in the caecum. If the animal is receiving an antibiotic, the synthesis of amino acids and vitamins will be retarded. Animals suffering from coccidiosis respond to supplemental pyridoxine (vitamin B_6) feeding.

Minerals most likely to be deficient are salt (NaCl), calcium (Ca) and phosphorus (P). Salt should be added to make up 0.5% of the total ration. Salt fortified with the essential trace minerals is often fed to ensure against deficiencies. If a complete mixed ration is not fed free-choice, salt (as in a salt block) should be available. Dietary requirements for growth are 0.5% Ca and 0.3% P, and for lactation 1.1% Ca and 0.8% P. Rabbits can tolerate high calcium diets by excreting the excess in the urine. Lucerne supplies adequate calcium for the diet, but if grass hay and grain are fed, a calcium supplement (such as steamed bone meal or calcium carbonate) should be included. High calcium intake is especially important for lactating does since so much calcium is lost in the milk. Calcinosis (a deposit of calcium salts in the tissues) may occur in Angora rabbits fed high calcium or wide calcium:phosphorus ratio diets or a ration containing more than 1500 IU of vitamin D/kg of feed. The preferred Ca:P ratio is 1.5:1.

A potassium dietary level of 3% is suggested for maximum digestibility of crude fibre. A copper level of 200 to 250 ppm in the diet may reduce enteritis, parasite infestation and danger from calcinosis and produces a generally tonic effect. Over 500 ppm copper, however, is toxic to the rabbit and the faeces will not be suitable for the growth of worms.

A rule of thumb states that a New Zealand doe and litter of 8 will consume about 50 kg of feed for the 12-week period from breeding to weaning the kits at 2 kg. Depending on their size, weanling rabbits eat 110 to 170 g of feed/day. The doe requires 3.5 to 4 kg of feed to produce a kg of gain in her suckling litter. The kits after being weaned require only 2.5 to 3 kg of feed to produce 1 kg of gain. This suggests the advantage of weaning as soon as the kits are consuming sufficient dry matter (3 to 6 weeks of age). About 7 kg of feed will be required to produce a slaughter weight of 2 kg, depending on the energy and protein concentration in the feed.

In the tropics grasses may appear lush, but yet have relatively limited feeding value for rabbits due to their low digestibility and low protein content. Tropical legumes, on the other hand, are much more digestible and can better serve as feed sources. Satisfactory performance is possible on a diet based on legume and grass forages with table scraps, kitchen wastes, vegetable tops, weeds, tree leaves and crop residues such as surplus or damaged vegetables and fruits such as bananas, plantains and mangos. During the dry season, when forage availability is low, re-breeding of the does can be delayed to reduce their nutrient requirements. Forages that remain green during the dry season might be planted to supply roughage. The optimal growth objective of fryers is to reach 2 kg at 8 weeks; only about half of this growth rate can usually be achieved on fresh forage alone.

Rabbit managers should be aware that there are some plant materials that contain toxic factors. Some toxins may have been produced by the plant itself (e.g. gossypol in cottonseed, trypsin inhibitor in soybeans, alkaloids, saponins, oxalates and goitrogens in other plants). In other cases toxins can be produced by organisms living in the feed material; to prevent this, care must be taken to keep feeds dry in storage. Moisture will allow undesirable bacteria and mould growth which may produce toxins, such as aflatoxins. If care is exercised, small amounts of some of these contaminated feeds, in combination with more suitable feeds, can often be tolerated by the animal, but since animals differ in their susceptibility, each case must be considered on its own merits.

Clean water should be available to all animals at all times. Restricting the water intake reduces the feed intake. With an environmental temperature of about 20°C the water intake will be 2 to 2.5 times the dry matter intake. Raising the temperature to 30°C increases water consumption by 50%. Lactation increases the water requirement threefold. Succulence in the feed will reduce the free water requirement somewhat. After 3 weeks, nursing kits require free water as they begin consuming dry feed. It is estimated that a meat type doe and her litter will drink 4 litres of water daily.

MANAGEMENT

Diseases

Problems associated with disease are minor when only a few animals are grown in any one place. These problems become compounded, however, when more animals are concentrated in small areas or are grown in the same area continuously over successive generations. The diseases to which rabbits are exposed will vary from place to place. However, special attention should be given to the following:

Pasteurellosis is caused by extremely contagious *Pasteurella multocida* and is manifested as snuffles. From sneezing and nasal discharge the disease may lead to pneumonia, abscesses, excessive lacrimation (tear formation), conjunctivitis, pyometra (uterine infection), orchitis (testicular infection) and wry neck. Tetracycline (100 mg per litre of drinking water) is recommended for controlling an outbreak.

Enteric diseases are associated both with the nature of the diet and with how it interacts with the environment such as the climate and season of the year. Fibre is especially important in this regard. Enlargement of, reduced motility and impaction of the caecum, enteritis, diarrhoea and enterotoxaemia occur when fibre levels drop to about 10%.

Enteritis is due principally to enterotoxaemia resulting from endotoxins produced by the growth of *Clostridium spiroforme*, *C. perfringens*, and *Escherichia coli*. The incidence is increased when diets low in fibre and high in energy are fed, leading to a 'carbohydrate overload of the hindgut'. It appears that *C. Spiroforme* produces toxin only when growing rapidly, and it grows rapidly only when there is present a readily available carbohydrate. Symptoms include profuse diarrhoea, reduced feed intake and rough haircoat. The condition is seen most commonly in 4 to 8 week old fryers that usually die 12 to 24 hours after signs of enteritis first appear. Feeding diets containing 15% to 18% crude fibre and low levels (250 ppm in the diet) of copper sulphate (blue vitriol) have some preventive value. In pet rabbits a rather common form of enteritis is a carrot-induced enterotoxaemia (the 'Bugs Bunny' syndrome).

Mucoid enteritis is recognised by the excretion of a jelly or vaseline-like material, consumption of large quantities of water and grinding of teeth. Rabbits produce a 'sloshing' sound when moved about. It may result from consumption of finely ground feed or of diets containing excessive levels (over 22%) of crude fibre. These lead to an impaction generally at the junction of the caecum and small intestine.

Coccidiosis can result from infection by several species of the genus *Eimeria*. Oocysts passed out in the faeces require moisture and warmth to form spores and become infective. More commonly in rabbits coccidia infect the bile ducts of the liver and shed oocytes in the faeces. Young rabbits 2 to 3 weeks old are most susceptible, frequently becoming infected by eating sporulated oocytes from their mother's faeces. The more oocytes ingested the larger the liver becomes and the slower the weight gain. Coccidiosis is controlled by maintaining rabbits in hutches with bottoms that allow the faeces to fall through and by preventing faecal contamination of feed and water. If a clinical outbreak appears, administer sulphaquinoxaline at less than 0.04% in the drinking water for 2 weeks, or add it to the feed at 0.025% for 3 weeks. High levels prove toxic to the kidneys. Other coccidiostatic drugs might be used instead.

Sore hocks are seen as a loss of hair on the foot pads and big sores on the hocks. Some rabbits are genetically more susceptible than others; they usually lack dense fur on the foot pad itself and have a small foot size. A rough cage floor can cause the problem. Animals developing sore hocks should be culled.

Hair chewing is a vice associated with stress (overcrowding, boredom, malnutrition, especially insufficient fibre or protein) as are the vices of cannibalism in chickens and tail biting in pigs.

Ear and skin mange are caused by microscopic mites. They are very common and cause severe losses in production. Inflammation, abscess formation and scabbing, usually along the face, ears and upper back, are typical signs. Affected rabbits scratch themselves continually (Figure 34). Early detection and treatment of mange generally insures rapid recovery (as is the case with most diseases). Ten drops of kerosene added to a cup (250 cc) of red palm oil provides a local remedy; one application of several drops on the affected areas is usually sufficient. In herds experiencing sporadic outbreaks of ear mange, bi-monthly preventive applications in the ears of all rabbits over 5 months of age will generally keep the problem under control. A number of more modern preparatory miticide medications are also available for ear mites.

Figure 34   Rabbit with ear mites.

Mastitis (blue breasts) is common in commercial rabbitries, and is occasionally seen in smaller rabbit production units. It attacks lactating does, especially where overcrowding and poor management practices including inadequate sanitation are encountered. Staphylococci, and occasionally streptococci, as well as other microorganisms are incriminated as causative organisms. The infected mammary glands become red, swollen, hot and painful, later turning blue (cyanotic). The doe will not eat, has high fever (40.6°C) and may die quickly. Penicillin, streptomycin and other antibiotics are often effective if injected early enough in the course of the disease.

Staphylococcus septicaemia (caused by *S. aureus*) is a highly contagious and highly fatal disease in some rabbitries. Young rabbits in particular are susceptible to acute outbreaks. Abscesses and suppurative infection of the skin, internal organs and mammary glands of older animals are common manifestations. Penicillin and other antibiotics may be effective in chronic cases; gentamicin and kanamycin are given by injection only.

Ringworm (dermatophytosis) is a fungus infection of the skin (contagious among most mammals, including man). Ringworm is characterised by circular, raised, reddened and flaky skin lesions, usually starting on the head and spreading to other areas of the body. Poor housing, poor sanitation, overcrowding and poor nutrition predispose an animal to this condition. Tinctures of tannic acid, salicylic acid and benzoic acid are usually effective and are relatively inexpensive. Sodium caprylate, salicresin, iodine, betadyne and tinactin are other effective skin medicines. Griseofulvin orally in doses of 25 mg/kg of body weight, or in the feed at the rate of 825 mg/kg of feed for 14 days is effective. (Effectiveness is improved if the oral medicine is accompanied by application of the above medicines on the skin lesions.)

Salmonellosis (caused by *Salmonella typhimurium* and *S. enteritidis*) is an uncommon disease of rabbits, but it can be very devastating and is on the increase throughout the world. It is characterised by yellow diarrhoea, septicaemia and rapid death. Abortion is common in pregnant does. Young rabbits and pregnant does are most susceptible, particularly when under stress. Transmission is by direct contact as well as from contaminated feed and water, bedding, faeces, kitchen waste, and infected animals and birds. Treatment is generally not effective; infected animals are best eliminated and good hygienic measures followed. However, some outbreaks may be controlled by adding 10 mg oxytetracycline per litre of drinking water for 10 days.

Other infectious diseases   A listing of other serious but not common infectious diseases that might be encountered would include: viral papillomatosis, pyometra (pus infection of the uterus), bite wound abscesses, pneumonia, conjunctivitis, hutchburn (urine burn), tuberculosis, listeriosis, rabbit pox, myxomatosis, spirochaetosis and nematosis. Tularaemia is found in some wild (cottontail) populations and can possibly be introduced into domestic rabbitries.

Parasitic diseases, other than coccidiosis, are not common in rabbits.

Larval tapeworms can be transmitted from dog faecal contamination of forage eaten by rabbits. The tapeworms are consumed and cysts form in the internal organs of the rabbits. The cycle is completed when the dogs eat the gastro-intestinal tracts of infested rabbits. The cycle can be interrupted by cooking the entire rabbit carcass before feeding to dogs or by keeping the rabbits off the ground in hutches.

Non-infectious conditions and injuries The rabbit grower might also encounter the following conditions that are not caused by disease micro-organisms: broken backs, hair chewing, cannibalism, heat exhaustion, moist dermatitis, dystocia, dental malocclusions, ketosis, plant poisoning, toxaemia, hair balls and congenital hydrocephalus.

Sanitation Practices

These should include periodic (at least weekly) removal of manure and other waste materials before ammonia levels can build up. Sprinkle manure beds with lime or super phosphate to reduce odour and flies. This treatment, at the same time, also improves the fertiliser value of the droppings.

Clean and sanitise pens, kindling boxes and other equipment before re-use. Use a stiff brush for hard-to-remove hair, manure and other debris. Thirty ml of sodium hypochlorite (laundry bleach, clorox) dissolved in a litre of water will serve as an effective disinfectant for cages and equipment. Also effective are Roccal, Betadyne and Nolvasan.

Another procedure is to use a 2% lye solution (1 g of lye to 50 ml of water) to sanitise cages and equipment. Beware that lye is caustic, and corrosive to aluminum, cotton and animal tissues.

Animals should be carefully observed for external parasites, abscesses or lumps, scouring, physical injury, listlessness, etc. Any animal showing signs of contagious disease should be eliminated or placed in an isolation cage and should receive treatment until full health returns. To prevent spreading the disease, treat any sick animals only after all others are cared for, then wash and disinfect hands and aprons carefully before coming in contact with other animals.

Response to Climatic Stress

Rabbits, if dry and protected from draughts, can withstand cold weather. They are much more capable of maintaining homeostasis at colder than warmer temperatures. They must be protected from rain, direct sunlight and high temperatures (above 30°C or 85°F). The comfort zone for rabbits is 15°C to 20°C (60°F to 65°F). In colder weather rabbits will eat more; increased dry matter intake will correspondingly increase water consumption. In warmer weather rabbits eat less. Heavily pregnant does and new-born litters are especially susceptible to heat stress. Heat stressed does are wet around the mouth with occasional light haemorrhage around the nostrils; they are restless and breathe rapidly. The newly kindled kits become extremely nervous whenever they are heat stressed.

As temperatures increase, it is generally considered that vasodilation

169

begins at about 20°C (68°F), panting at 35°C (95°F) and death at 44°C (100°F) when the body temperature reaches that temperature. Increased humidity increases the heat stress in animals. Does nearing parturition and rabbits with respiratory diseases are more susceptible to the ill effects of heat stress.

Heat stressed animals should be placed in a quiet, well-ventilated place with a wet surface such as wet feed sacks on which to lie and cool off. Immersing the rabbit in cool water for 3 seconds is an emergency measure. However, wet coats are predisposing causes of pneumonia and respiratory ills. Bucks might be made infertile by exposure to 30°C (85°F) for 4 to 5 days. They retain libido but fail to serve the does for about 60 days after such exposure. Respiratory stress on rabbits is compounded by relative humidity above 35% and by ammonia from urine and faeces. These conditions are promoted by the accumulation of manure under the pens and by inadequate drying and removal of moisture from the pen. Air quality involving relative humidity, air movement (draughts) and concentration of noxious gas (e.g. ammonia) is more important to the rabbit than is the air temperature. Higher temperatures can be tolerated by animals if during the night temperatures drop, giving the animal some respite.

Buildings and Equipment

Shelter can be constructed from locally available building materials that discourage gnawing. Enclosed buildings should be avoided where possible. Buildings should protect rabbits from any factors which would interfere with their comfort or performance such as weather, sunshine, noise, natural enemies, snakes, thieves, dogs, predators or other frightening situations. Does will trample or eat their young if sufficiently excited. Buildings, pens and dropping boards (if used) should be designed for convenience to allow for good cleaning, moisture and manure removal and disinfection. Generally, within each hour there should be about 20 air changes. Draughts should be avoided. Economy in expense and labour should be a goal.

Rabbits should be caged about 1 m off the ground. Cages can be satisfactorily constructed of many materials including bamboo and raffia palm. However, many prefer to use 2.5 × 5 cm 14 gauge galvanised mesh wire for the sides and top and 1.2 × 2.5 cm on the bottom. Dirty hutch floors are a source of major health problems such as sore hocks, coccidiosis and other diseases associated with filth. If bottoms are made of wire, a piece of board or other flat surface should be provided for the animal to rest upon, reducing the likelihood of developing sore hocks (Figure 35, p. 172). Wood sticks, raffia palm or bamboo, which provide better support for the legs and feet than wire, also reduce the incidence of sore hocks. If such materials are used in construction, the floor openings should be spaced at 1.2 to 2 cm to allow faecal pellets and urine to pass through. Fryers can be kept in colonies but should be provided with a minimum of 464 sq cm per animal. If pens are outside they should be portable to allow moving them to clean ground as needed. Breeding bucks and does and Angoras should be caged singly. After sexual maturity at 12 to 14 weeks the rabbits should be separated into male and female groups, then placed in individual cages by 16 weeks of age.

Buck cages can be round to make easier natural mating. If several bucks are maintained, they should be separated to prevent physical contact and fighting, especially in the presence of a doe. Buck cages should be centrally located. Surrounding areas may need protection from spraying urine. The cage should have doors that open easily and all four corners should be easily reached.

Cage sizes vary with the animal size and function. The following dimensions should be provided as a minimum:

	Surface m²	Width cm	Depth cm	Height cm
Breeding animals	0.36	60	60	40
Growing animals up to 2.7 kg weight each	0.05	—	—	25
Angora rabbits	0.36	50	60	45

Does may need more space if nursing kits beyond 4 weeks of age.

Rapid urine drainage is necessary. When relatively high temperatures prevail urine evaporates rather rapidly. If proper ventilation is provided, the faecal pellets need not be removed from under the cages at daily intervals.

Under some situations with coccidia control in place, rabbits can be grown in colonies. The floor should be paved and sloping for easy cleaning. Each doe should be provided with her own kindling box. Three or 4 does that have been raised together can be grouped together; introducing a strange doe will result in fighting. At breeding time only a single buck can be in the pen at one time; he should be removed after 3 or 4 days. The does will kindle in their own boxes. The number of kits can be equally distributed among the nursing does. A lack of retrieving instinct makes this fostering possible.

Yet another system has been suggested for hot, dry areas. A vertical-sided pit about 1 m deep and 2 m long is dug. One or 2 does with their water and feed are added. Shade must be provided until the rabbits have burrowed holes to get out of the heat. These facilities cannot be used indefinitely because of diseases and an inability to control the animals in their burrows.

Another possibility for protection from hot (or cold) weather would be an underground box lined with stone or concrete to discourage digging and escape. A passageway to the above ground hutch can be made from 5.1 × 30.5 cm boards. To make it easier for the animals to climb, small wooden cleats can be nailed across the floor every 30 cm. This underground space could serve as a nesting box in hot weather. The box should have access from above for periodic inspection.

Hay racks can reduce waste and/or contamination of forage. Racks can be

Figure 35   Protective shade, fresh water and a solid resting surface.

made from sticks 30 cm tall spaced about 4 cm apart, horizontally joined at top and bottom, and mounted to one corner of the cage. Suspending tied bundles of forage from the top of the cage is an alternative to constructing hay racks. Locate feeders so they can be filled without opening the cage door. A screened bottom will allow the 'fines' to fall through and not accumulate.

Many objects can be used as feeders and waterers such as split bamboo sections, clay, ceramic or cement-cast bowls, empty tins and even glass bottles inverted into a small, shallow tin, such as an oblong sardine can. In any case, containers must be tied securely to the cage to avoid waste and spills. Water must be available at all times. The water requirement is increased both in hot and cold weather.

A conveniently located table or counter should be provided for working with the animals (weighing, identifying, checking for pregnancy, etc.).

Herd Records

Records of all breedings, kindlings, weights and weaning dates should be maintained. This is a very important management practice necessary for

172

good animal care and for culling the non-productive animals and selecting replacement stock for genetic improvement.

Each animal in the breeding herd should be identified and records kept of:

For the doe:

1. date does were bred;
2. date and conclusion from pregnancy test;
3. date of kindling;
4. whether a good nest was provided and if kindling occurred normally in the kindling box;
5. number born (both dead and alive)
6. number of young weaned;
7. litter weight at 21 and 56 days of age.

For the buck:

1. date buck was used and whether pregnancy resulted;
2. number kindled from each successful mating;
3. number weaned per litter;
4. litter weight at 56 days of age.

Handling Tips and Summary

1. Toenails may grow too long and require trimming when rabbits are confined in hutches. Hold the rabbit's foot up to the light to observe the translucent cone. Cut off the tip 2 mm beyond the cone with sharp, strong clippers. This may need to be done twice a year.
2. Keep hutches clean and sanitised.
3. Provide fresh water every day, shade and a solid surface on which to rest (Figure 35).
4. Wash and sanitise water containers each week, at least.
5. Feed rabbits away from direct sunlight and at least once daily, preferably in the evening.
6. Keep feed troughs clean and free from mouldy, spoiled feed. The feed also should be kept dry.
7. For the breeding female adequate nutritional supply is especially important during the following periods:
   a. One week after mating to assure low embryonic mortality;
   b. One week before kindling because 85% to 90% of the foetal growth occurs at this time.
   c. Three weeks following kindling to support maximum milk yield and growth of the kits who depend on maternal milk as sole feed during their first 2 weeks.
8. For mating, put the doe in the buck's hutch and remove her when mating is completed.
9. Check doe for pregnancy 12 to 14 days after mating and re-breed if not pregnant.
10. Put nest box in the hutch 27 to 28 days after mating (3 to 4 days before kindling).

11. Leave all kits with the doe for 48 hours; then reduce the litter to the number of teats possessed by the doe. If other does have kindled recently with less kits it may be possible to have them foster the excess, hence the advantage of breeding 2 or more does on or near the same day.
12. The quantity of milk produced by the doe increases for the first 21 days, after which it declines. This decline corresponds with the amounts of dry feed eaten by the kits.
13. Provide enough feed to keep breeding animals in a thrifty condition. Increase the feed if they become thin, decrease the feed if they become fat.
14. Feed young growing rabbits and lactating does all they will clean up.
15. Watch for and regularly treat any cases of ear mites. Eliminate from the herd any animals with sore hocks.
16. Always pick up rabbits by the fold of skin over the shoulders and support the hind quarters with the other hand. Smaller animals can be lifted at the hips (Figure 36). Do not lift rabbits by the ears.

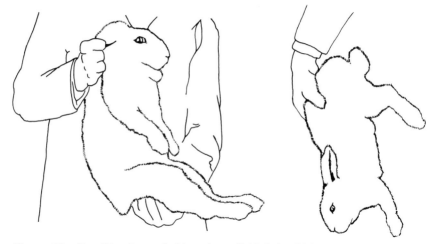

Figure 36   Handling large (left) and small (right) rabbits.

17. Provide adequate ventilation but protect rabbits from draughts and dampness.
18. Keep records on each breeding animal. They can be identified by tattooing the ear.
19. Practise sanitation. Keep the hutches and area clean and sanitised.
20. Disease prevention is better than cure.
21. Diseases are readily transmitted when moving animals from one rabbitry to another. Quarantine new animals for 21 days prior to mixing them with others.
22. Protect rabbits from disturbances especially before, during and after kindling. Avoid unnecessary handling of kits for several days after birth.

23. Rabbits require fibre in their diet but do not digest it well.
24. Male and female rabbits should be separated before 14 weeks of age. After 16 weeks of age males and females should be kept in individual cages. Males can be castrated to reduce fighting.
25. Only animals of the same age should be caged together.
26. Human contact and handling have no ill effects on young rabbits.
27. Sudden, intense noises are harmful to the well-being of rabbits. This is especially true with does having young kits.

Slaughter

Slaughter yields vary from 54% to 65% depending on age, breed, sex, body fill or shrink, and whether the head, heart, lungs, liver, feet and skin are included with the carcass. Heavier slaughter weights produce higher slaughter yields with higher meat to bone ratios, even though it takes increasingly additional energy and protein to produce that weight. Before puberty the increase in body weight consists primarily of protein tissue. Weight gain in older animals is primarily fat. Due to the inverse relationship between fat and water, the ratio of carcass protein is barely influenced by feeding intensity and age.

When given a complete free choice feed, the fryers will have desired fat development when they have achieved about 60% of the adult weight for the breed. With only roughage feeding, rabbits will achieve 80% of their mature weight before achieving the same amount of fat development.

OTHER USES FOR RABBITS

Rabbits are primarily kept for meat so the foregoing information pertains especially to meat production. However, rabbits also serve as sources of wool and pelts.

Wool and Hair

Angora rabbits reportedly only need 30% of the digestible energy of sheep and 90% of Angora goats to produce the same amount of wool. In Angora rabbits the wool fibres grow 0.6 to 0.8 mm daily up to a length of 20 cm or more with the wool having about the same diameter as hair on other rabbits. Wool is obtained by shearing 4 to 5 times per year or by plucking the moulting hair up to 8 times per year. Graded wool is at least 6 cm long. Does produce more wool than bucks. Annual yields of up to 1 kg of wool can reasonably be expected in well-managed rabbitries in temperate climates; less has been reported in other areas. The Angora not only has an increased hair growth, but is also more docile and easy to shear.

Wool production has a demand on the nutrient supply which makes Angora breeds less efficient than other breeds in producing meat. Concentrate feeding can increase wool production 10% to 20% over roughage feeding alone. The greatest yield is obtained at 10 to 14 months of age, and animals remain in production for 3 to 4 years.

Raising the temperature from 18°C to 30°C decreases the wool production 14% but also decreases the feed requirement by 31%.

Angora rabbit wool production is a labour-intensive operation requiring management skills and a developed marketing and manufacturing system to be profitable.

If true, the beneficial soothing effect on painful joints and muscles that comes from contact with Angora wool fabrics should be good news to arthritics. If the advantages in meat and wool production over red meat and other wool producing animals are real, rabbit production should experience accelerated growth especially where labour is relatively abundant and cheap.

Pelts

Rabbit pelts have been largely replaced in the furrier trade by the partially domesticated mink, fox, nutria (coypu), raccoon, etc. However, some improved varieties of Rex rabbits have superior pelts and colours. These animals are larger and meatier than older varieties, being truly dual purpose; while providing excellent pelts they also yield satisfactory carcasses for meat.

Pelts of older (over 6 month) rabbits are preferred to 11-week-old fryers. The best pelts come from rabbits raised singly in outside cages with straw bedding and slaughtered at the beginning of winter. Pelts of bucks tend to be thicker and harder to tan. If thin-skinned pelts are an important product, young males should be castrated at the time their testicles emerge at about 3 to 4 months of age. To reduce bleeding the arteries in the spermatic cord should be scraped or squeezed off.

Tanning can be accomplished with either inorganic chemicals or the more traditional vegetable materials. If the hair is not wanted, it can be removed and the hide can be used to manufacture leather products.

Defleshing and Cleaning Skins for Tanning

The success and ease of tanning skins for fur or leather, and the quality of the product obtained, depends to a large extent on the defleshing and preparation of the skin. This is especially true for rabbit skins, which are usually quite thin, easily torn, and in some cases actually delicate.

Defleshing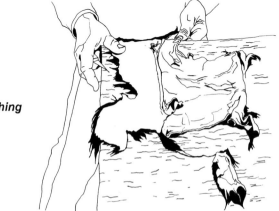

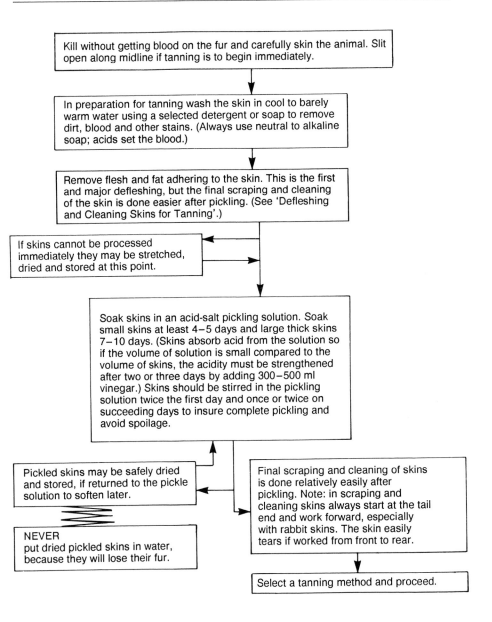

Kill without getting blood on the fur and carefully skin the animal. Slit open along midline if tanning is to begin immediately.

In preparation for tanning wash the skin in cool to barely warm water using a selected detergent or soap to remove dirt, blood and other stains. (Always use neutral to alkaline soap; acids set the blood.)

Remove flesh and fat adhering to the skin. This is the first and major defleshing, but the final scraping and cleaning of the skin is done easier after pickling. (See 'Defleshing and Cleaning Skins for Tanning'.)

If skins cannot be processed immediately they may be stretched, dried and stored at this point.

Soak skins in an acid-salt pickling solution. Soak small skins at least 4–5 days and large thick skins 7–10 days. (Skins absorb acid from the solution so if the volume of solution is small compared to the volume of skins, the acidity must be strengthened after two or three days by adding 300–500 ml vinegar.) Skins should be stirred in the pickling solution twice the first day and once or twice on succeeding days to insure complete pickling and avoid spoilage.

Pickled skins may be safely dried and stored, if returned to the pickle solution to soften later.

NEVER put dried pickled skins in water, because they will lose their fur.

Final scraping and cleaning of skins is done relatively easily after pickling. Note: in scraping and cleaning skins always start at the tail end and work forward, especially with rabbit skins. The skin easily tears if worked from front to rear.

Select a tanning method and proceed.

Figure 37 *General outline of steps for cleaning and pickling small furs.*

Excellent results can be obtained in defleshing and cleaning skins by spreading them on a flat firm surface covered with 2 layers of blanket or a piece of carpet having pile of uniform height. Spread skin on the soft surface, flesh side up, and clamped on one side or one end with a 2.5 × 5.1 cm (1 × 2 in) board and C-clamps.

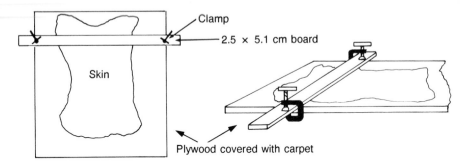

A very good and perhaps preferable method is to hinge one end of a 2.5 × 5.1 cm board to one side of the cleaning surface so that it presses down on the soft surface. The other end is free to be lifted up and then clamped down tightly to firmly hold one side or one end of the skin.

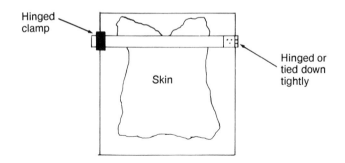

After a fresh skin is washed to remove blood, dirt or other material that may stain the fur, the skin is clamped in 4 appropriate places so that the large pieces of fat or flesh, if any, can be carefully scraped off. Exercise care to avoid tearing the skin.

A most important point is that this scraping or cleaning should always be done working from the tail end to the head end of the skin.

In the case of rabbit skins, the thin layers of flesh adhering to the skins should be left on until after the skin is pickled, at which time they can be removed much more easily and safely.

After rabbit skins have been in a pickle solution 5 or 6 days they are usually ready to be cleaned and tanned. The final cleaning should be done very carefully, otherwise the skin will be torn or badly damaged.

Clamp the skin on the cleaning surface with the fur against the soft surface.

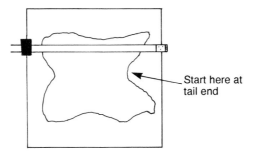

Starting at the tail end of the skin, very carefully separate the surface membrane and flesh from the skin. Be very patient, gradually separating and pulling the 2 layers apart.

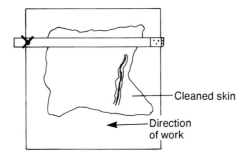

When about 5 cm or more of skin has been cleaned on the lower half of the skin, turn the skin around so that the rest of the rear part of the skin can be cleaned.

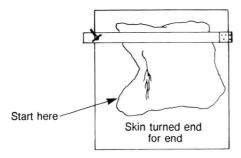

Continue separating the fleshy membrane from the skin of the tail end. When 5 cm or more of skin has been cleaned entirely across the tail end of the pelt, release the skin and reclamp it with the cleaned portion placed under the 2.5 × 5.1 cm holder board.

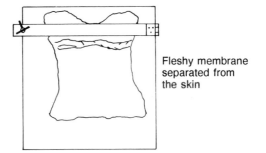

Fleshy membrane
separated from
the skin

At this point of progress the job can be completed by gently pulling down on the fleshy membrane and carefully scraping with a spoon, blade or moderately sharp edge. Special knives are available for scraping and cleaning skins. Its shape gives a sharpened arc to work with and makes cleaning or scraping skins easier and safer.

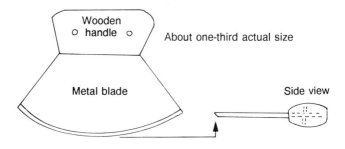

Wooden
o handle o   About one-third actual size

Metal blade   Side view

The blade edge is bevelled on one side only. If the blade is held by the handle in the palm of the hand with the thumb on the side with a cutting edge and fingers behind on the bevelled side, the blade can be used to clean skins with the sharp edge parallel to or away from the skin. This minimises skin cuts. When scraping skins, if the blade is held in the same way but perpendicular to the skin, down strokes will lead with the cutting edge and be very effective.

Generally a good policy to follow in cleaning skins is to leave a little membrane or fleshy material rather than tear up the skin trying to clean up a very difficult area.

In the final finishing of a tanned skin small areas of fleshy membrane on a skin can be removed by carefully sanding with medium to fine sandpaper. Large areas that are poorly cleaned, however, result in uneven and poor tanning. An added note, if a 0.1 N NaOH (4.1 g of lye dissolved in 1 litre of water) is used to remove the hair from a goat or deer skin, a hardness of the surface grain of the leather will result.

Pickling

After the skins have been scraped and cleaned of adhering flesh and fat, they can be either stretched and dried for storage for later processing or they can be pickled. The pickling process entails soaking in a salt and acid solution to start the preservation process. The 2 most common sources of acid are vinegar and muriatic acid. Vinegar is a common household product, entirely safe to use. On the other hand, muriatic acid is a commercial source of concentrated hydrochloric acid (HCl). Due to its strength, HCl must be handled with caution. If any should come in contact with the skin, it should immediately be washed away with much water or a solution of sodium bicarbonate (baking soda).

The pickling process begins with preparing the acid–salt solution. Two pickling formulae that have proved to be successful are:

1. Using a container that will hold at least 4 litres, combine:
   2.6 litres of water
   625 ml of vinegar
   450 g of salt

2. Or, if concentrated hydrochloric (muriatic) acid (HCl) is used (concentrated HCl is 28% to 31% acid) prepare pickling solution:
   3 litres of water
   45 ml of HCl
   450 g of salt

In a container large enough, cover the skins with the pickling solution and soak small skins 4 to 5 days; large, thick skins require soaking for 7 to 10 days. During this time the skins absorb the acid from the solution and since the amount of solution is small in relation to the skins, the acidity must be strengthened after 2 or 3 days by adding 300 to 500 ml of vinegar. The first day the skins should be stirred twice and during each succeeding day stirred once or twice. This stirring ensures complete pickling and retards spoilage.

At this stage the process can be interrupted by drying and storing. However, to reinstate the process, skins must be resoaked in pickling solution until completely soft. (N.B. Use pickling solution because skins resoaked in water will lose their fur.)

After pickling a final scraping and cleaning can be performed rather easily. The same precautions as observed in the initial scraping must be applied to the pickled skins. Any of the several tanning methods can be used.

Methods of Tanning

Salt acid bath and plant extracts Skin is tanned with tannic acid as in tea leaves, oak leaves, etc. using the following steps:

1. Remove the skin from the pickle solution, squeeze dry and spread out in the tanning solution, flesh side down. Leave in the solution 3 days. Each day lift the skin and stir the tea before replacing the skin.

2. Prepare the tea:
   2.8 litres of water
   230 g of tea or oak leaves containing high levels of tannic acid
   750 ml of vinegar
   570 g of salt

   Bring the water to a boil, add the tea leaves and simmer for about 30 minutes. After the strong tea has cooled somewhat, add the salt and vinegar to it. Pour the tea (tea leaves and all) into a container such as a plastic lined cardboard box about 10 cm deep with an area about 50 × 60 cm. Press and stretch out the skin in the tea.
3. After 3 days give the skin a quick rinse to remove tea leaves and stretch it on a board flesh side out and allow to dry. After drying, work and rub the skin by hand until soft.
4. Thoroughly wash the skin in a mild vinegar solution (1 part vinegar to 16 parts water) to remove excess salt.
5. Re-stretch and re-dry the skin, working it during the final phase of drying. Continue working (massaging) it by hand until it becomes pliable and soft.

This is about the most simple tanning method to be found. The tanning agent, tea leaves, can be replaced by the leaves and twigs from any plant species having a reasonably high tannic acid content. In any geographical area there will no doubt be many species containing sufficient tannic acid for good tanning; as an example, the whole oak family would be good. Tannic acid in plants can be detected by a puckering sensation when leaves or stems are chewed.

One negative aspect of the tannic acid process is a slight staining (or dyeing) by the tea. White furs appear a 'creamy' colour. This staining can be avoided by the following procedure:

1. Make a soap gel by dissolving 28 g of soap in 500 ml of hot water and allow it to cool.
2. Take a pickled skin and thoroughly clean the flesh side removing flesh, fat and non-skin membrane.
3. The pickled, cleaned skin is spread out flat, fur side up on a large, flat surface. The fur is then rubbed with the soap gel until all the fur fibres are covered with soap. The skin is then put into the tanning solution and spread out, flesh side down.

The acid of the tanning solution immediately precipitates the fatty acid of the soap as an insoluble film around the fur hairs protecting them from the staining action of the tea.

After 3 or 4 days, when the skin has become completely 'tanned' by the tea, it is given a quick rinse with water to remove the tea leaves, some of the salt and the acid from the fur and then thoroughly washed in a neutral or very slightly acid detergent solution.

This washing will remove the fatty acid from the fur and leave the skin slightly acid with only a little salt remaining. It should then be stretched and dried, fur side up.

Note: washing a 'tea tanned' skin in alkaline solution removes some of the tea. When the skin is nearly dry (just slightly damp) it can be worked (manipulated) a little by hand and worked again once or twice while drying completely. It will be finished as a nice soft skin without any fur stains.

Traditional American Indian brain tissue tan Prepare the tanning material as follows: mix 500 ml of brain tissue with 250 ml of water in a small pan and heat to boiling; simmer for about 20 minutes. The mix is then 'homogenised' (stirred vigorously) as in a blender for about 1 minute. This same result could be obtained by vigorous stirring with a hand egg beater or even a 'whisk'. Twist the moist pickled skin in a folded towel to remove excess moisture and stretch and pin it on a cardboard (fur side down). Paint the brain suspension mixture on the flesh side of the skin leaving a fairly thick layer. Place the cardboard and skin in a large plastic bag to prevent drying, and set it aside. (The American Indians used this method on 'de-haired' skins and after covering both sides of the skin, rolled it up and covered it with something, and then set it aside for 1 to 2 days.)

After about 12 hours uncover the skin and paint it again with the 'brain mix' and set aside in the plastic bag again. Repeat this repainting after 24 hours. After 36 hours uncover the skin and paint it with water and allow it to dry uncovered. When the skin is completely dry, rub and work it until soft. (The tanning material that remains appears as a waxy residue on the surface which falls off as the skin is worked.)

After the dry skin is softened wash it thoroughly in water and squeeze it dry. Now rinse in a solution of Epsom salts (30 ml in 2 litres of water). Stretch the skin flat and dry to a mildly damp condition. Then stretch and work the skin intermittently until completely dry.

N.B. The 36 hours treatment and washing loosens a lot of the fur.

Following the tanning of a skin and final drying, a careful scraping of the tanned skin when it is placed over a soft surface can greatly enhance the softness. The specially shaped knife referred to previously can be used effectively for softening tanned skins in this way. The skin should be worked from the centre toward and across the edges.

USES FOR MANURE

Rabbit faecal pellets from fryers are 60% to 70% dry matter and from breeding does 40% to 50%. Green feed reduces these dry matter values. Annual production is estimated at 30 kg per fryer equivalent and 60 to 80 kg per breeding doe. These pellets can be used as conditioners and plant nutrient in the soil or can be used as animal feed. Up to 10% dried faeces can be added to the rabbit diet without affecting the feed conversion. Rabbit pellets can make up to 60% of the pig diet. Care must be exercised that diseases are not transmitted through faecal feeding.

Dung as Feedstuff for Vermiculture

Manure from rabbits and other animals (including chickens, if diluted) is suitable as a feed for earthworms such as *Eisenia* (also called *Helodrilus*) *foetida* (manure worm), and *Lumbricus rubellus* (red worm or red wiggler, so named because they squirm when handled). The African nightcrawler is also popular where freezing does not occur. The night crawler (*Lumbricus terristris*) is not commonly raised commercially because of its slow reproductive rate and special requirements. Worms can produce up to 200 times their initial weight each year.

Characteristics of Earthworms

Worms are most active at 14°C to 27°C (60°F to 80°F). Freezing will kill the worms but they will continue to breed up to 40°C (100°F) if given adequate moisture.

The growth medium for worms should contain 15% to 30% moisture, being crumbly moist, not soggy wet. If the soil is too dry, worms will either burrow deeper into the soil, hibernate with a subsequent loss of weight, or they will die. Moisture in the medium can be reduced and aeration fostered by adding some (5%) straw, leaves or dried manure.

The earthworm has a well-developed muscular, nervous, circulatory, digestive, excretory and reproductive system. The earthworm is divided into about 90 to 150 segments; different biological functions are performed in different segmented sections.

Earthworms derive their energy and protein from the organic matter contained in the soil that is eaten. Feed enters the digestive tube of the earthworm where both enzymatic and fermentative digestion occur. Nutrients are removed and the undigested material is excreted as 'castings'. The presence of 1% to 2% added dirt or sand to manure, by providing grit for the gizzard, will enhance digestion. The organic matter consumed by the worm should contain 9% to 15% protein (N x 6.25); a commercial earthworm feed contains 19% protein. A feed with insufficient protein will not support growth of worms, and too much protein increases the organic matter decay rate resulting in heating. Also high protein encourages house fly and other maggots as well as ammonia odour to develop. In earthworms kidney-like excretory organs secrete to the exterior slimy, nitrogen-containing waste that lubricates the surface of the worm and at the same time stabilises the burrows and casts.

Worms grow 6 to 8 times more on steer and sheep droppings than on oat straw. A mature worm can ingest 16 to 20 g of dry matter each year representing 11% to 14% of its body weight daily. Compare this with other animals that consume only 4% to 5% of their body weight each day. The ratio of converting the dry matter of rabbit faeces into earthworms is about 2:1. One preliminary test showed 10 kg of raw cow faeces (15% protein dry matter) being converted to 1 kg of earthworms.

Earthworms thrive best in a near-neutral medium with a pH of 6 to 7.2. The addition of lime to manure, while preventing the development of more acidity, can also reduce odours and supply calcium. This will also increase the soil fertilising value of the residue.

Earthworms lack specialised breathing organs. Respiration occurs by diffusion of oxygen and carbon dioxide through the moist body surface. Worms will not drown but can live as long as 31 to 50 weeks in aerated water. They will generally leave their burrows when the carbon dioxide concentration exceeds 20% to 30%.

Reproduction of Earthworms

Worms are bisexual, having both male and female reproductive organs. Although they can be self-mating, usually two worms mate and mutually exchange sperm. Mature sperm and ova together with a nutritive fluid are deposited in a capsule (cocoon) produced by the conspicuous central section (clitellum) of the worm. This capsule then slips off the worm and is deposited in the soil. Under optimal breeding and growth conditions, a mature earthworm will produce an egg capsule every 7 to 10 days or 36 to 52 capsules per year. These eggs have the appearance, size and colour of a grain of rice. The egg capsule will hatch in 2 to 3 weeks into 2 to 20 (average 7) young earthworms. With proper feed and care these young will mature to breeding age in 2 to 3 months. Up to an additional year is required to reach the full maturity length of 3 to 4 inches. Thus, it is possible for one mature breeding worm to produce 1200 to 1500 offspring or 10,000 descendants annually. The lifespan of an earthworm is not precisely known but it is doubtful that it will exceed 2 years.

Managing Earthworms

Earthworms can be considered as a form of livestock having certain minimum requirements of care and scheduling. Worm beds should be protected from sun and rain and extreme temperatures. Boxes of any convenient dimensions can be prepared with a depth of 20 to 30 cm generally preferred; the worms generally occupy only the top 15 cm. A solid under floor (e.g. concrete or galvanised iron) will prevent worms from wandering into the ground.

Prepare a bed by covering the floor with a 5 to 10 cm layer of organic material such as dried, broken-up manure, leaves, seed pods, bark or sawdust that has been dampened (15% to 30% moisture). If boxes are placed under rabbit cages, urine might be harmful to the worms and should be separated if possible. If this is not possible, or if the box is situated under chicken cages, twice each year the beds should be emptied and renewed because an acid condition develops from the faeces and urine which retards the growth of worms. Each week the beds should be stirred with a rake. Excess moisture and wet spots should be avoided. Drainage can be accomplished by covering the floor with a layer of gravel before laying down the bed.

To seed, add worms together with some of their original bedding material to the box. Some suggest seeding with 2000 to 4000 adult worms per square metre. One doe and litter should be able to support 1000 worms. If the bed is not situated under cages, after one week, it should be covered with an additional 5 to 10 cm of accumulated faeces. After 3 to 4 months the worms should be ready to harvest. Thereafter, worm numbers should be reduced monthly, or as needed, to keep the beds from becoming overpopulated. To harvest earthworms, skim off about 10 cm of material from the top of the bed and place it on a sunlit surface through which the worms cannot burrow. The bright light will drive them downward, the top layers of bedding can then be skimmed off leaving the worms exposed. They can then be picked out for use. Commercial machines have been developed for harvesting worms. Under optimal conditions, 100 g of worm mass can be produced from 1 kg of rabbit faeces.

Cold, heat, too little or too much moisture and mould formation are damaging to worm numbers. Other enemies are ants, springtails, centipedes, slugs, mites, certain beetle and fly larvae and other insects, birds, rats, snakes, moles, gophers, toads and yet other animals that might feed on worms. In addition, the earthworms might be plagued with various harmful protozoa and nematodes.

In closed buildings the suitability of placing beds under rabbit cages has been questioned. This is because working the bed to harvest the worms releases ammonia and other undesirable gases. These gases, together with the high humidity that is favourable for worm production, increase the rabbits' susceptibility to respiratory diseases.

Earthworm Uses

Earthworms average 23% dry matter and 58% protein and 2.8% fat on a dry basis. They can serve as an excellent source of protein for chickens, pigs, rabbits and fish, or when properly prepared, for humans. *Ver de terre*, French for earthworm, is the term applied to earthworms for human food. To remove residual soil and destroy undesirable bacteria, *ver de terre*, like *escargots* (snails), must be first washed in cold water then boiled. Earthworms have reportedly served as food in Japan and South Africa, and for the Maoris in New Zealand.

Worms reduce the formation of ammonia and other odours and flies in accumulated faeces. They also reduce the total faecal mass that must be removed from pens and cages. The castings and bed residues make excellent potting soil for horticultural plants.

10
RABBIT LEARNING ACTIVITIES

I BUILDING A RABBIT HUTCH

Purpose A hutch should be constructed in order to:

1. Confine the animal so it can be handled and cared for.
2. Protect the rabbit from predators—dogs, snakes, etc.
3. Shelter the rabbit from rain, wind, direct sunlight and heat while giving ventilation
4. Provide the animal with facilities for feed and water and a place for does to kindle.

Procedure

1. Choose appropriate construction material available in the region. Material should be resistant to gnawing.
2. Sketch a double hutch showing dimensions (for ideas see Figure 38 and Figure 1, p. 7). Sketch a feed trough and water source. Sketch a removable kindling box.

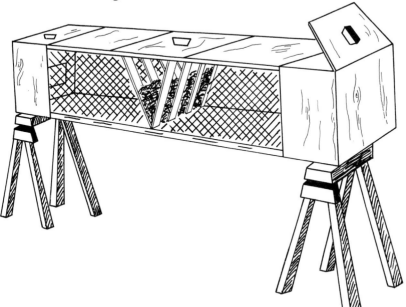

Figure 38   Constructing a rabbit hutch.

3. Locate hutches in the shade where they will be protected from the heat of the sun.
4. Observe ways of preventing losses by thieves and predators, such as:
   a. locate the rabbitry near the living quarters;
   b. enclose the rabbitry with a fence;
   c. place locks on the cage doors;
   d. attach bells, chimes, etc. to cages to notify when the cages are opened;
   e. train any family dog to protect the rabbits.

Check your learning Can you:

● State the functions that a rabbit hutch should serve, or why build a hutch?
● Sketch a double hutch for breeding does using materials available in the area. Include facilities for feeding, watering and kindling.
● Using this sketch as a pattern, build a hutch.

II SELECTING RABBITS FOR BREEDING

Purpose

1. Choose a breed.
2. For foundation stock select animals that are healthy, and have desirable conformation.

Procedure

1. Learn the principal body parts of a rabbit.
2. Choose a breed that produces meat and has done well in the area. New Zealand (white, red or black) or Californian (white with coloured ears, nose and feet) are of medium size (3 to 5 kg) and suitable in most circumstances.
3. Obtain rabbits from a reputable grower who maintains a disease-free herd. Notice whether premises are kept clean and the rabbits free from disease.
4. Start with junior does and bucks 2 to 3 months old so the animals can become adjusted to the new circumstance and in order to acquaint the new owner with rabbit production.
5. Choose only healthy animals. Avoid animals with snuffles, diarrhoea, ear canker, sore hocks, ear mites and other diseases and pests. Avoid also genetic defects such as wry tail or face, crooked feet and legs, lop ears and buck teeth (malocclusion).
6. Choose rabbits with desirable conformation (Figure 39):

   head—broad, symmetrical (not pear shaped), evenly carried ears;
   back—wide and long providing a large back and loin for meat;
   hips—wide and rounded and long, standing slightly above the back level;
   legs—provide the principal source of meat; should be straight, neither cow-hocked nor pigeon-toed;

188

A. Desirable conformation. Well balanced throughout, tapering from front to rear, full rear quarters with well-rounded rump, and well-carried ears.

B. Narrow head and shoulders, weak in midsection and loin, cut off hips with rough protruding hip bones, side-carried dewlap and belled ears.

C. Beefy in shoulders, full, wide rump but flat over top of rear quarters, cow-hocked (legs bowed outward), and open-carried ears.

Figure 39 Rabbit conformation. (Sketches after ARBA.)

189

belly—filled in but not pot-bellied; shows 8 or more teats; body length—long and smooth.

While these are standard considerations of the desired appearance of rabbits, it should be recognised that there is no scientific evidence to show that one conformation is more efficient than another in producing meat.

7. Breeding stock should show rapid growth and come from fast growing parents.
8. Breeding animals should have the ability to reproduce regularly and possess sound temperaments with calm behaviours.

Check your learning Can you:

• Choose a breed of rabbit and state the reasons for the choice.
• Identify one or more sources of foundation stock.
• Supply the names of the different parts of a rabbit as shown on a sketch of a rabbit.
• Describe 3 or 4 ways in which an abnormal animal differs from a healthy, normal one.
• When shown sketches of different conformation types, identify types that would be the more desirable.

III MATING RABBITS

Purpose Learn how to mate a buck and doe.

1. Become acquainted with the external genitalia by examining these on live male and female rabbits. Study the internal reproductive structures from sketches and/or from rabbits that have been butchered.
2. Practise handling rabbits. Lift the animal by the skin over the shoulder while providing support under the rump with the other hand.
3. The age of does and bucks at their first mating depends on the breed and size of the animal, size being more important than age. Generally, does should be 5 to 6 months of age and bucks 1 month older.
4. Put the doe in the buck's hutch (not vice versa).
5. Recognise the behaviour upon mating. Copulation should occur almost immediately, after which the buck falls off the doe. If rabbits have not mated after 5 minutes, return the doe to her hutch and try again tomorrow. In hot weather, breed during the early morning hours.
6. If mating occurs, return the doe to her own hutch and record the date and identity of both the buck and the doe.
7. If does are in good shape and adequate feedstuffs are available, return does to the buck after 18 days postpartum.

Check your leaning Can you:

• Identify the different parts of a rabbit's genitalia (male and female) from a diagram.

- Properly pick up, handle and pacify a rabbit.
- State the age and size at which young rabbits can be bred.
- Explain why the doe should be placed in the buck's hutch for mating, and not vice versa.
- Describe how to recognise when a successful mating has taken place.
- Prepare a chart indicating how the mating information should be recorded for the herd records.

IV DETERMINING PREGNANCY IN RABBITS BY PALPATION

Purpose Learn how to palpate a doe for pregnancy.

Do not wait until the 31–day gestation period has passed to learn if a mating resulted in pregnancy or pseudopregnancy; 12 to 14 days after putting the doe in with the buck, any developing embryos can be felt through the abdominal wall (Figure 30, p. 153).

To conduct the pregnancy test, place the doe on a non-slip surface and allow her to relax because tight abdominal muscles can make diagnosis difficult. She will be easier to handle and less frightened if you back her into a corner of her own hutch. Restrain her by holding the ears and fold of skin over the shoulders with one hand and place the free hand under the body slightly in front of the pelvis; place the thumb on one side of the abdomen and the fingers on the other; exerting light pressure, move the thumb and fingers gently back and forward. Any embryos will be felt as marble-sized forms.

Some skill is involved in the test; any does diagnosed non-pregnant (open) by the inexperienced operator should be re-palpated a week later. With the development of proper techniques, pregnancy can be determined by the tenth day. In early testing, to distinguish between the embryos and faecal pellets, remember that the uterus lies at the bottom of the abdominal cavity (ventral) whereas the large intestine is above the uterus nearer the backbone (dorsal). If properly done, no harm can come to either the doe or foeti from conducting this pregnancy test before day 15.

If the test has been properly conducted and no embryos are detected, the doe should be returned to the buck for re-mating.

Mature bred does when diagnosed pregnant should be switched from a maintenance ration to a pregnancy ration. If she has not kindled by day 34 after breeding, she should be put back with the buck.

Check your learning Can you:

- Recognise the pregnant doe as distinguished from the open one. Palpate for pregnancy 2 or more does, some open (non-pregnant) and some known to be 10 to 14 days pregnant.
- Explain how the embryos felt inside the doe. How can they be distinguished from faecal pellets?

191

V   KINDLING AND LITTER CARE

Purpose   Learn necessary steps in kindling and caring for the litter.

The gestation period for rabbits is about 31 days. Place a clean nesting box in the pregnant doe's hutch 27 days after breeding. A satisfactory nest box will provide seclusion for the doe and comfort and protection for the litter. In cold weather especially, the nest box should contain some short chopped straw, dry, fine-stemmed grass hay, shavings or other bedding; the doe will pull hair from her hip, dewlap and mammary areas to complete the nest.

Kindling usually occurs at night. If disturbed or frightened, a doe may drop her kits all around the hutch and/or eat them.

The day after kindling when the doe has left the box, quietly remove any dead or abnormal kits and leave no more than 8 kits, or the number of teats on the doe. Extras can be fostered to other lactating does with similar aged (1 to 2 days) offspring. Full stomachs in the kits indicate a nursing doe; her milking ability can be assessed by the 21-day total weight of her litter. The litter weight should increase seven- to eightfold during the time it is in the box. Avoid unnecessary handling of the kits for a few days after birth. Does may abandon newborn young that have been handled.

The box should be inspected periodically to replace soiled bedding. The young should open their eyes 10 to 11 days after birth. The doe eats the kit's faeces but the urine either drains from the box or is absorbed into the bedding. Remove the box from the cage 2 to 3 weeks (or as early as possible) after kindling. The boxes should then be thoroughly cleaned and disinfected before re-use.

After 28 days, lactation declines. Under optimal conditions, the kits will be eating enough solid feed so they might be weaned by removing the doe. Then after 2 or 3 days the young rabbits should be placed in growing hutches for 4 more weeks at which time, if adequately fed and managed, they are ready to market. The doe should be returned to her pen.

If adequate feed is not available or if other conditions are not optimal and the doe was not re-bred at 14 to 18 days, the kits might stay with the doe for up to 8 weeks at which time they should be fryer size.

Any animals kept for breeding stock should be sexed before 8 weeks. Holding the rabbit by the ears and shoulder skin with one hand, rest it on its rump with its feet in the air. With the free hand press on either side of the exterior genitalia. If a doe, the sex organ will appear as a slit with a slight depression at the end next to the rectum; the buck displays a raised circle through which the penis might be made to protrude.

Check your learning   Can you:

- Describe how a kindling box should be prepared for the doe.
- State the number of days after breeding that the kindling box should be put in the hutch of the pregnant doe.
- Recite the usual behaviour of a doe that has been disturbed at kindling.

- Specify the anatomical and physiological circumstances that limit the number of newborn kits a doe can handle.
- Give the age of kits at which the nesting box should be removed from the pen.
- Give the age at which kits are weaned.
- Detail the factors influencing the age at which rabbits become fryers. How might these factors be influenced by genetics?
- With live rabbits, distinguish between a male and female by examining the genitalia.

VI NUTRITION AND FEEDSTUFFS FOR RABBITS

Purpose Become familiar with the rabbit's digestive system, the nutrients required and the feedstuffs to be fed to supply the nutrients.

Digestive System

The parts of the digestive system (see Figure 11) and the functions they serve are:

1. Mouth—with its teeth, lips, cheeks and tongue to secure and chew feed.
2. Oesophagus—a tube to transport feed from mouth to stomach.
3. Stomach—an enlargement to store feed and begin protein digestion.
4. Liver—the largest organ in the body serving to produce bile to aid in fat digestion and absorption, store nutrients and produce many metabolites.
5. Pancreas—to produce enzymes for the digestion of fats, carbohydrates and proteins; also serves as an endocrine gland producing several hormones.
6. Small intestine—a convoluted tube to produce enzymes and absorb the products of digestion.
7. Caecum—a site for fermentation of fibre and absorption of fermentative by-products.
8. Large intestine—a sacculated tube that separates the coarser, more fibrous parts of the ingesta forming round faecal pellets. It also absorbs water.
9. Anus—a sphincter muscle closing off the intestinal tract.

Rabbits are simple stomached herbivores consuming most types of greens, hays and grains. An enlarged caecum and large intestine (colon) provide a place for a limited amount of fermentation to occur. These organs produce two kinds of faecal matter. The soft fraction, retained by the caecum, is composed of the more digestible components of the feed and is re-eaten (coprophagy) by the rabbit directly as it passes from the anus to pass again through the digestive system. The more fibrous fraction of the faeces is expelled in a dry pellet form. This separation process, together with the ability to select and eat only the more digestible parts of the plant, allows the rabbit to thrive on plant materials without carrying around a large mass of fermenting digesta. It also allows survival on a herbivorous diet with a

193

crude fibre digestibility of only about 15% (compared with 44% in goats). Caecal and intestinal fermentation provides enough volatile fatty acids to meet 10% to 12% of the rabbit's daily energy requirement. Coprophagy largely frees the adult rabbit from a dependence on dietary sources of vitamins B and K and to a lesser extent the essential amino acids.

The diet must be palatable to the rabbit for optimal results. If a feed is not acceptable, a great deal is wasted as it is scratched out of the feeders by the animal searching for more palatable portions. Enteric diseases seem to be associated with the nature of the diet and with environmental interactions.

Nutrients

The nutrients required by the rabbit and the function these nutrients serve are much the same as in other animals and man.

Carbohydrates (sugars, starches, fibre) are the primary and cheapest source of energy (measured in kilocalories). In addition, the rabbit has a fibrosity requirement that can be supplied by a minimum of 12% crude fibre in the diet. For best results the grain should be fed with or after the roughage; otherwise impaction might develop near the mouth of the caecum, or the caecum and hind gut could become overloaded with starch leading to fermentative production of enterotoxins and of enteritis. Leaves and especially stems of plants supply fibre as well as protein and other nutrients.

Fats are a concentrated form of energy. While 6% to 10% is optimal, fat can be fed to provide up to 20% of the diet without ill effects. The level of fat fed will depend largely on the relative energy costs and availability of fats and grains.

Proteins are necessary to supply the amino acids for growth and maintenance of life. If there is insufficient available carbohydrates and fats to meet the body energy needs, proteins can be burned to supply the same energy equivalent as do carbohydrates. With coprophagy, the bacterial synthesis of protein in the caecum and large intestine makes a contribution to the essential amino acid requirement of the mature but not the very young rabbit. Even the mature rabbit on usual rations benefits from feed supplements of argenine, lysine and methionine. The rabbit is capable of utilising efficiently the protein in forage plants. Minimum dietary crude protein levels for growth, maintenance, pregnancy and lactation are 16%, 12%, 15% and 17%, respectively, assuming a balanced essential amino acid supply for the young.

Vitamins are dietary factors essential for the various processes of life (metabolism). The water-soluble vitamins are the B-complex (thiamin, riboflavin, niacin, pyridoxine, pantothenic acid, biotin, folic acid, cobalamin, choline) and vitamin C (required in the diet only of man, monkeys, guinea pigs and some fruit-eating birds). While much of the needed B vitamin requirement is formed in the gut, not enough is synthesised to free the rabbit entirely from a dietary requirement. The fat-soluble vitamins are vitamins A, D, E and K. Vitamin K is synthesised sufficiently in the gut. Green, sun-cured roughage can meet these other requirements.

Minerals fall into 2 categories, the macro elements and micro elements, depending on the relative amounts required by the body.

The macro elements to be included in the rabbit diet are: calcium, phosphorus, magnesium, sodium, potassium, chlorine and sulphur. The micro elements are: iron, copper, cobalt, manganese, zinc, molybdenum, chromium, vanadium, selenium, silicon, iodine, fluorine. A mixed diet that will supply the needed energy will generally also supply the minerals required. A notable exception is salt (NaCl). Small amounts of various trace elements are frequently added to it to ensure against any deficiencies, and then it is fed free-choice.

Water is essential for life. It should be clean and always available.

Feedstuffs

The composition of some feedstuffs is shown in Table 2 and Figure 16. The primary plant feedstuffs are:

1. Indigenous plants—native forages collected in the 'wild' state from the bush or fields. These include grasses, legumes, forbs and weeds, as examples.
2. Cultivated forage—improved varieties collected from cultivated plots for livestock feeding.
3. Crop residues—plant aftermath usually following harvest, e.g. leaves from beans, maize and peanuts, sweet potato vines and cassava cuttings.
4. Agricultural by-products—consisting of processed wastes from grains, fruits, nuts, roots, tubers, vegetables, etc. Examples include wheat bran, rice bran, brewers' dried grains and sugar cane.
5. Food wastes—food unsuitable for human consumption, such as stale bread and tortillas, damaged and/or surplus fruits and vegetables, peelings and parts of roots and tubers, maize husks, peanut hulls and table refuse.

Essential dietary nutrients are found in various feedstuffs:

1. Protein—legumes, oilseed meals, fish meal and processed cereal grains.
2. Energy—plant vegetative growth, fruits, vegetables, cereal grains, roots, tubers, table wastes, vegetable and animal oils.
3. Fibre—grasses, legumes, weeds, forbs, garden wastes, kitchen refuse.
4. Minerals and vitamins—trace mineralised salt, common salt, bone meal, wood ash, vitamin premix, oilseed by-products (rich in P), and succulent roughages (provide vitamin A).
5. Water—clean, fresh and available at all times.

Roughages

These can be cut and fed fresh (green chop or soilage) or ensiled or dried and fed as hay. Care should be taken to preserve the leaves because leaves contain the greater proportion of digestible energy, protein, vitamins and minerals. Leaves are also the most readily available parts of plants to be consumed by rabbits.

Legumes (lucerne, clover, etc.) are more palatable and are richer in protein, vitamin A and calcium than are the grasses. Tropical grasses have a lower protein and higher fibre content than grasses grown in temperate zones. For the highest yield of digestible nutrients, forages should be harvested when immature, just before they reach the reproductive (boot or flowering) stage.

Other roughages provide essential fibre but are of poor digestibility and acceptability. Examples are straw and chaff, corn cobs, cottonseed hulls, bagasse and wood. Dry roughages may be ground and, where equipment is available, incorporated into a pellet for feeding.

In small-scale rabbit raising, green feeds can be fed free choice. It should be recognised that green feeds contain more moisture and, therefore, 4 to 5 times as much must be eaten in order to get the same amount of dry matter as in dry hay. In fact it is usually impossible for the animal to consume enough green feeds alone to meet the nutrient requirements for maximum production. Harvested green feeds should not be left in piles because if they become heated before feeding, they might cause digestive disturbances.

In addition to lucerne, clover and grasses, succulent feeds might include comfrey, vegetables and/or vegetable tops, amaranths, roots and tubers, and weeds such as dandelions. Kikuyu (an African grass) grown in Central America from stolens (no seeds) can also be fed.

Roughages provide the indigestible fibre (fibrosity) needed to prevent enteritis.

Concentrates

Energy concentrate sources include the grains such as oats, barley, wheat, maize, sorghum and sunflower seeds. Which grain is used will depend on its cost per unit of energy and/or protein, and acceptability by the rabbit. The energy comes principally from starch. Of the cereals oats and barley have the highest fibre content and are preferred by rabbits; barley and wheat have the highest protein. All grains provide little or no calcium, but are fair sources of phosphorus. Grains should be coarsely ground for optimal acceptability and digestibility. Milling by-products such as wheat bran and beet pulp are palatable and rich in fibre, protein and phosphorus and can be fed to provide up to 25% of the entire ration. Molasses is about 50% sugar and can be added at the rate of up to 3% to 10% of the diet to increase palatability and reduce dustiness.

Protein sources are concentrates that contain 20% or more of protein. The residues remaining after extracting the oil from oil-bearing seeds are rich in protein and phosphorus. Since high quality lucerne meal often contains over 20% protein it could also be considered a protein supplement.

Soybean meal, when economically available, is the preferred protein supplement since it is palatable and provides a good balance of amino acids. Rabbits grow equally well on raw soybeans as they do on a commercial soybean meal as a protein source. The trypsin inhibitor in raw soybeans does not appear to produce harmful results in rabbits.

Cottonseed meal is usually cheaper and more available than soybean meal, but is less palatable and is deficient in the amino acids lysine as well as methionine. Cottonseed meal contains toxic gossypol, the effect of which can be reduced by adding available iron (ferrous sulphate) to the diet. Five per cent of the diet is considered a safe maximum for feeding cottonseed meal to rabbits unless the meal has been de-gossipolised.

Groundnut meal is equivalent to soybean meal. It, like other concentrates, when warm and humid can serve as feed for the mould *Aspergillus flavus* which produces aflatoxin, a strong carcinogen.

Check your learning Can you:

1. Sketch and label the rabbit digestive tract.
2. List by name the 6 classes of nutrients.
3. Name one or more functions that each nutrient serves in life.
4. Give an example of each of the following feedstuffs that are locally available: roughage, energy concentrate, protein concentrate.

VII FEEDING RABBITS

Purpose Learn how to develop a ration suitable for maintenance, growth, pregnancy and lactation.

Table 2 supplies composition values of some common feedstuffs. Table 10 and Table 11 contain data after NRC 'Nutrient Requirements of Rabbits', 1977. Mature rabbits can adjust their feed intake to compensate for energy levels in diets that vary from about 2000 to 3000 kcal DE/kg, so a specific energy concentration (kcal/kg) in the diet is of less consequence. A ration containing 2100 kcal DE/kg is satisfactory for maintenance of mature rabbits. Diets containing 2500 kcal DE/kg and 16% to 18% crude protein are generally satisfactory for all types of rabbits.

Table 10 Nutrient Requirements of Rabbits Fed Ad Libitum (percentage or per kg of diet)

Nutrient	Maintenance	Growth	Gestation	Lactation
Digestible energy (kcal/kg)	2100	2500	2500	2500
TDN (%)	55	65	58	70
Crude fibre (%)	14	10–12	10–12	10–12
Crude protein (%)	12	16	15	17

Rabbits satisfactorily consume many kinds of feedstuffs. They prefer the more fibrous grains of oats and barley over wheat, maize and milo (sorghum). Grains supply energy and, when accompanied by a protein supplement like soybean or cottonseed meal, they provide adequate protein and phosphorus as well.

Table 11   Examples of Adequate Rabbit Diets

Kind of Animal	Ingredients	%	Total Diet Digestible Energy kcal/kg	Daily Intake g
Maintenance, does and bucks, 4.5 kg	legume hay barley grain salt	70 29.5 0.5	 2200	150
Growth 0.5 to 4 kg	lucerne hay maize grain barley grain wheat bran soybean meal salt	50 23.5 11 5 10 0.5	 2400	50 to 205
Pregnant does and working bucks 4.5 kg	lucerne hay barley grain soybean meal salt	50 45.5 4 0.5	 2500	185
Lactating does 4.5 kg	lucerne hay wheat grain sorghum grain soybean meal salt	40 25 22.5 12 0.5	 2600	235

When fed at 20% of the ration, good quality sun-cured lucerne hay will supply adequate vitamins A and D and calcium and most other vitamins and minerals. Mature animals can thrive on good quality sun-cured lucerne hay alone, but maximum production requires additional energy. Sufficient vitamin K and B-complex are synthesised in and absorbed from the lower gut to meet the needs of mature rabbits.

Salt (NaCl) should make up 0.5% of the diet. This demand can be met by adding a salt block to the hutch for free choice consumption. The disadvantages of the salt block are greater cost, greater labour requirements and cage corrosion.

Legume hays are preferred to grass hay because they are significantly more readily digested, higher in protein and calcium and are more palatable to the rabbit.

Grains should be coarsely ground; rabbits tend to develop diarrhoea when all the feed is finely ground.

Table wastes except meat, fat or spoiled foods are acceptable to rabbits. Many weeds can be fed depending on availability and rabbit acceptability. Wide and rapid changes in feed composition, especially those involving high moisture forages, should be avoided, otherwise digestive upset and diarrhoea can result.

Rules of Thumb

The following general rules and assumptions might be used in developing rations and determining quantities of feedstuffs needed for rabbits. Such general information is useful for budgeting or other purposes if more precise data are not available in tabular form.

Under optimal conditions, kits should weigh about 50 g at birth, then they should gain 300 g to weigh 350 g at weaning when 42 days of age. They should then gain 30 g per day in order to reach a slaughter weight of 2 kg after feeding for 8 weeks post weaning.

It is generally assumed that 45 kg of feed is required to feed a doe and her litter from kindling to weaning the kits at 8 weeks.

Feed conversion on a balanced diet should be 1 kg of gain for every 3.6 kg of feed. The maintenance diet should contain 2100 kcal DE/kg; the rations used for other needs should contain 2500 kcal DE/kg.

Rations fed to non-producing animals should be 16% to 22% crude fibre, while those fed to growing and producing animals should contain 14% to 16% fibre. The protein content should be 12% for maintenance, 15% for gestation, 16% for growth and 17% for lactation.

The specific nutrient requirements of rabbits have not been as well established as for the chicken and some other species. As far as energy is concerned, requirement values reported in the literature are based on a given energy concentration and percentage of protein in the feed. The assumption is then made that the rabbit will eat enough to meet his needs, whatever they are. It would be helpful if daily energy requirements were expressed in terms of units such as kcal of digestible energy (DE)/kg. Then, knowing the DE in various feedstuffs, the quantities needed could be better estimated and budgeted into the operational plans.

Thus, the energy concentration of the feed is set at a level at which the animal will be able to consume sufficient to meet its energy needs. The required nutrients are then set in the feed at such levels that when the animal eats enough to meet his energy requirements, he will at the same time consume the required amount of these nutrients.

Some factors which alter feed intake are stress, palatability, the variation in nutrient content of feedstuffs and the availability of these nutrients.

Meeting Nutrient Requirements

The composition of some of the common feedstuffs has been summarised in Table 2, pp. 44–45. Table 10 lists the nutrient requirements of rabbits when they are fed all they will eat. Examples of adequate diets have been listed in Table 11 as derived largely from the NRC publication 'Nutrient Requirements of Rabbits', 1977.

In designing a ration, similar feedstuffs can be substituted with minor adjustments. For example, if maize is substituted for barley, the energy content of the diet will be slightly increased and protein and fibre decreased. Replacing the lucerne hay with grass hay in a ration would require an increase in the protein and calcium content of the concentrate in order to maintain the same level of nutrient intake. Lucerne and clover are of similar composition when in the same stage of maturity and can be exchanged without further adjustment.

The proportions of a grain and a protein supplement needed to produce a concentrate supplement of a desired nutrient composition can be easily determined by using the Pearson square. As an example, mixing a supplement containing 16% protein for pregnant does using maize at 9% protein and soybean meal at 44% protein:

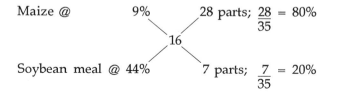

Thus, a mixture of 80% maize and 20% soybean meal would have a 16% protein content.

Successful rabbit production depends on adequately meeting their nutritive requirements. These requirements vary with such factors as age, size, breeding activity, pregnancy and lactation. Since rabbits are capable of widely adjusting their feed intake to meet their energy needs, generally rabbits of all types, sizes and physiological conditions perform well on a ration that contains about 2500 kcal digestible energy per kg and 16% to 18% crude protein. The lower the energy content of a diet, the lower would be the necessary protein and other nutrient levels. This is because the animal will eat more feed to meet its energy requirement. At the same time, the rabbit has a specific quantitative requirement of so many grams of protein or other nutrients. The more the animal eats in meeting its energy requirement, the more would be the nutrient intake. Consequently, each gram of a low energy feed would need to contain less protein in order for the rabbit to meet the specific protein requirement. More calcium and phosphorus is required for lactation (1.1% and 0.8%) than for growth (0.5% and 0.3%); less is needed for maintenance. Salt should make up 0.5% of the total diet.

Looking further into the feeding of rabbits, weanling rabbits normally grow very rapidly and will consume about 3 to 4 kg of feed per kg of gain (feed conversion rate).

Although the specific requirements for the various nutrients exist, they have not been widely reported in detail. As a practical matter, sound judgement should govern the feeding of animals. Frequently, this means supplying free choice the best roughages available together with limited feeding of any concentrates that might be on hand.

A 3 kg rabbit requires 244 kcal DE for maintenance. About 9.5 kcal of digestible energy is required per gram of body weight gain, regardless of the energy content of the diet; however, there is a limit to how much consumption can be adjusted, so if growing rabbits are to gain at a maximum rate they must have a diet containing at least 2500 kcal/kg.

Only during the last trimester of pregnancy will the nutrient requirement of pregnant does increase materially. The rule of thumb is to increase the ration from 1⅓ times maintenance at the beginning of pregnancy to double maintenance at pregnancy's end.

During lactation, assume that a 3.6 kg doe produces a modest 114 g of milk daily, this milk containing approximately 250 kcal. Assume further that 45% of the feed energy is converted to milk energy, so the feed energy requirement for lactation would be 550 kcal, which is almost double the maintenance requirement.

The maintenance requirement for a 3.6 kg rabbit is 330 kcal DE which is about 1.5 times the basal metabolic rate. Thus, the total digestible energy requirement for a 3.6 kg doe producing 114 g of milk would be 880 kcal. It is more efficient to feed the kits directly rather than indirectly through the lactation process. Therefore, kits should be weaned as soon as possible, usually between 4 and 6 weeks of age under good management conditions.

Referring to Table 2, pp. 44–45, 'Composition of Some Common Feeds', early bloom lucerne hay has 2430 kcal DE/kg. Then, 880 divided by 2430 would tell us that 0.36 kg of hay daily is necessary to meet the requirement.

If the feeder wanted to feed his doe equal parts of full bloom lucerne hay (2280 kcal/kg) and maize (3500 kcal/kg), 1 kg of the 50–50 mixture would contain 2280/2 = 1140 kcal from the hay and 3500/2 = 1750 kcal from the maize, for a total of 2890 kcal/kg. To meet the 880 kcal requirement, the doe would require 800/2890 = 0.277 kg of the mixture, or 138 g of hay plus 138 g of maize.

While these exercises provide general information on the quantities to feed, it should be recognised that feed composition values are only approximations and represent all of the feed, leaves, stems and all. In practical situations the rabbit will eat the leaves; most of the stems will remain.

Summarising, rabbit feeding suggestions would include:

1. Supply only fresh feed. Do not allow feed to accumulate and become contaminated with mould etc.
2. Offer a good variety of feeds daily.
3. Have feed, salt and water always readily available to the rabbits.
4. Do not overfeed high energy feeds.

Check your learning   Can you:

- Develop a rabbit ration for:
  a. maintenance;
  b. growth;
  c. gestation;
  d. lactation.
- Calculate the proportion (per cent) of barley and cottonseed meal to be mixed to yield supplement containing 18% protein for a growing rabbit.
- Specify the digestible energy and crude protein concentration in a satisfactory ration.

- Explain why rabbits can maintain productive levels on feed containing varying concentrations of energy.
- Solve the following:
  a. Using the information in Table 2, pp. 44–45, calculate the amount of full bloom clover hay (70% of the ration) and barley grain (30% of the ration) to feed to a doe requiring 700 kcal digestible energy:

  Lucerne 2280 kcal/kg × 70% = 1596 kcal
  Barley   3330 kcal/kg × 30% = 999 kcal

  Total   2595 kcal/kg
                               × 1/700 = 0.270 kg feed

  b. How much of a 50% lucerne hay, 50% maize ration would be needed to supply 800 kcal DE?
- Determine and itemise the quantities of roughage, concentrate and mineral supplement needed to maintain 4 does and 1 buck through a production cycle from one kindling to the next.

VIII   SANITATION AND HYGIENE

Purpose   To understand the need for sanitation and disease prevention among animals.

Diseases cause illness and death. Healthy animals are free from disease.

Look through the microscope and see some of the tiny forms of life (called microorganisms) that are present but cannot be seen with the unaided eye. Most of these microorganisms do not hurt us; in fact, many are beneficial. However, some of these organisms do make us sick (or cause diseases). Those forms of life that are harmful are called pathogens. They are of several different forms such as viruses, bacteria, moulds and protozoa. Diseases might also be caused by worms and insects.

Listing but a few examples of disease: pneumonia, snuffles, mucoid enteritis, mastitis, staphylococcus septicaemia and enterotoxaemia. Different localities have different diseases that are common to the area (endemic).

Pathogenic organisms are so small that they can be seen only with the aid of a microscope. In order for them to cause damage, the pathogens must first enter the body through the digestive tract, respiratory system, intact or injured skin, or other body openings. Pathogens are transferred from sick to healthy animals in manure, body secretions and coughing.

To prevent diseased animals from making the healthy ones sick too, the ailing animals should be kept separate from the healthy ones. When new animals are brought onto the farm or if older animals have been away and brought back, they should be kept separate for 3 weeks (called quarantine) so that if they had been infected and subsequently became ill they will not contaminate all others.

Wild birds and animals frequently carry disease organisms which can cause the disease in domestic animals. Therefore, wild birds and animals should be kept away from the feed, pens and domestic animals.

If the animals are well fed, protected and healthy they are more resistant to the pathogens. Animal resistance to some diseases can be increased to higher levels by vaccination.

In addition to invisible pathogens, internal and external parasites can also cause damage to animals and so must be controlled. These include protozoa, roundworms, tapeworms, flukes, mites, ticks and fleas. The purpose of sanitation is to destroy or weaken the pathogens before they can harm the animal. Chemicals that destroy pathogens are called disinfectants or sanitising agents.

Sanitation begins with cleanliness. Manure, rotting feed, hair and rubbish should not be allowed to collect, but should be either tilled into the land for fertiliser or burned.

Pens and cages in which sick animals have been kept should be thoroughly cleaned and disinfected. Also cages, nesting boxes, etc., should be cleaned and disinfected after each use.

Disinfectants can be prepared and applied to clean surfaces by several methods. Manure and other dirt can neutralise the effects of most disinfectant, because the disinfectant reacts with the dirt rather than the pathogen; therefore, the equipment should be cleaned before the disinfectant is applied. Disinfectants are more effective when applied hot. In addition to commercial preparations, some effective disinfectants are:

30 ml sodium hypochlorite (laundry bleach) dissolved in 1 litre water.
217 g lye (NaOH) in 1 litre of water.
40 ml cresol per litre of water (emulsification will be facilitated by the addition of a detergent).

When animals are kept confined on the same land for long periods of time or when large numbers of animals are concentrated in small areas, disease organisms and internal and external parasites build up in numbers and virulence. The pattern can be broken by moving animals to fresh uncontaminated grounds or buildings and by tilling the ground.

Summarising the steps in proper cage sanitation:

1. Prepare the chosen disinfectant solution.
2. Remove rabbits from cages.
3. With a dry brush, brush off the entire cage surface.
4. Disinfect the cage and leave undisturbed for 1 hour.
5. Rinse the cage well with clean water.
6. Once dry, replace rabbits in the cage.
7. Provide fresh feed and water.

Check your learning Can you:

- Name the forms of organisms that cause disease.
- What are some general symptoms shown by diseased animals?
- State how a disease is communicated from one animal to another.
- Explain how quarantine can help control disease.

- Describe why wild birds and animals should be kept away from domestic animals and their feed.
- Explain why equipment and surfaces must be cleaned before applying a disinfectant.
- Identify at least one disinfectant.
- Describe how to break the buildup cycle of pathogens and internal and external parasites.

IX DISEASES OF THE RABBIT

Purpose Learn the prevention, recognition, and treatment of metabolic and infectious diseases and internal and external parasites.

Disease is a condition that impairs the functioning of an animal. It can be due to the upset of metabolic processes or to pathogens or parasites. Disease control is important to rabbit production since mortality (death) from disease may be as high as 20% in many well-managed herds in the temperate zone. A greater loss should be expected in tropical areas. Although there are many diseases, certain ones more prevalent in some areas than others, the following are frequently encountered.

Enteritis—inflammation of the intestine—can result from several factors including enterotoxins. Dietary changes, or diets low in fibre and high in readily available carbohydrates, can upset the normal functioning of the gut, stimulating the growth of pathogens such as *Clostridium perfringens* (or *C. spiroforme*), *Escherichia coli* and rotaviruses which produce toxins. These enterotoxins not only damage the gut but are absorbed into the bloodstream, causing systemic damage and death. This condition is controlled by increasing the fibre in the feed. *C. spiroforme* produces toxin only when it has sufficient starch to support rapid growth.

Enterotoxins may be produced at levels harmless to an adult, lactating doe, but they can be picked up and secreted in the milk at sufficiently high levels to kill the kits. Staphylococci in the mammary gland can cause mastitis, which can produce sufficient toxins to kill the lactating doe suddenly, especially those with their first or second litters.

Finely ground feed, lack of roughage or excessive high fibre (over 22%) can produce impaction in the intestine at the juncture of the caecum. The animals drink large quantities of water which produces a sloshing sound when handled. Animals also grind their teeth. Affected animals excrete a jelly or vaseline-like material which leads to the condition termed mucoid enteritis. No treatment is known.

Coccidiosis affects both the intestine and the liver. The intestinal involvement produces diarrhoea, poor appetite, weight loss and sometimes death. Coccidia enhance the effects of enteritis. Liver coccidiosis produces a clogged bile duct and disfigured liver. Coccidiosis, tapeworms, stomach worms and other internal parasites can best be controlled by keeping the feed and water free from contamination with manure. This can be accomplished by keeping the rabbits in wire bottomed hutches which allow the faeces to drop through.

Ear and skin mange (canker) is caused by a mite that is very contagious being readily transferred from one animal to another. Two to 3 weeks after being attacked by the mite, inflammation and severe irritation forming yellow and brown scabs appear in the inside of the ear. The rabbit scratches its ear and shakes its head constantly.

Treatment consists of using a swab dipped in hydrogen peroxide to remove the scabs and then applying to the affected areas any number of materials such as a proprietary product, mineral oil, mineral oil containing 1% phenol, or any benzyl benzoate preparation. To prevent spread, burn all swabs, bedding and other materials that came in contact with the infected rabbits and thoroughly clean and disinfect the cages.

Snuffles result from a *Pasteurella* bacterial infection and are seen as a thick, sticky white discharge from the nose, which the animal wipes with his forelegs, together with constant sneezing. Secondary infections of pneumonia, pleurisy or haemorrhagic septicaemia can cause death. Prevention lies in good nutrition and ventilation and the immediate elimination of all affected animals.

Abscesses due to *Staphylococcus aureus* and/or *Pasteurella multocida* occur under the skin and elsewhere requiring the disposal of the affected animals and cleaning and disinfecting the premises.

Check your learning   Can you complete the table?

Disease	Symptoms	Treatment and/or control
Enterotoxaemia		
Mucoid enteritis		
Coccidiosis		
Ear mange (canker)		
Ringworm		
Snuffles		
Pneumonia		
Mastitis		

X SLAUGHTER AND PREPARATION OF MEAT

Purpose   Learn to kill, skin, eviscerate and cut up a rabbit.

Rabbits can be butchered for meat when weighing 2 to 3 kg, which under good conditions can occur when 8 to 10 weeks old.

1. Kill the animal by holding it by the hind legs with the left hand. Place the right thumb on the back of the neck and fingers around the throat.

Quickly snap the neck by pulling straight downward (Figure 40). Or, while holding the rabbit by the hind legs, strike it sharply behind the ears on the base of the skull with a heavy object.

2. Hang the animal by passing a hook or tying a cord between the tendon and bone of the right hock.
3. Cut off the head and let the carcass bleed.
4. Cut off the tail and the left hind foot at the hock joint, and the front feet at the elbow.
5. Slit the skin just below the hock of the right (suspended) leg, inserting the knife under the skin on the inside of this leg and cutting to the root of the tail; continue with the other hock.
6. Separate the edges of the skin from the carcass and pull the skin down over and off the animal, separating the fat from the skin with the knife.
7. Unless the skin is to be tanned within a few hours, while the skin is still warm, put it on a stretcher with the fur inside. This stretcher can be made in many ways, one of the simplest being a piece of heavy stiff wire (Figure 41). The wire can be removed as soon as the pelt has completely dried.
8. Make an opening slit in the carcass from the tail through the breast bone along the midline of the abdominal wall.

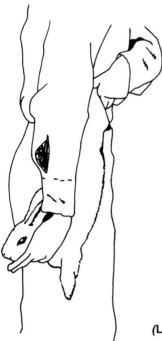

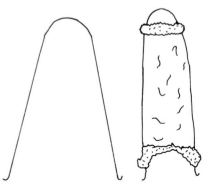

A stiff wire is bent and slipped inside the pelt. The tension of the wire to straighten out will stretch the skin.

(Above) Figure 41   Stretching a rabbit pelt.

(Left) Figure 40   Procedure for killing a rabbit.

206

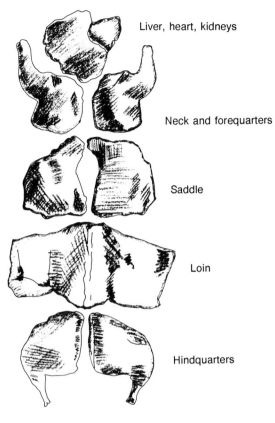

Liver, heart, kidneys

Neck and forequarters

Saddle

Loin

Hindquarters

Figure 42   Parts of a rabbit carcass.

9. Remove the viscera.
10. Remove the eviscerated carcass from the hook and rinse it with cold water to remove hair, blood or other unwanted materials. Dry off with clean cloth or paper towel.
11. Cut the carcass into 2 hind legs, the back, 2 rib halves, 2 front legs and the neck (Figure 42). The liver, heart and kidneys should be salvaged.

Check your learning   Can you:

- Kill a rabbit.
- Skin a rabbit.
- Place a skin on a stretching wire.
- Eviscerate a rabbit.
- Cut up a rabbit into 8 pieces.

XI   TANNING THE RABBIT PELT

Purpose   Learn to tan pelts so they can be used for clothing or bedding for the family or for sale.

A simple tanning method described by Daniel O. Robinson makes use of products readily available in the home—salt, vinegar and leaves containing tannic acid. The first is a 'pickling' process using vinegar, salt and water. The fat and flesh should then be scraped from the inside of the skin. Of the various tanning methods, tannic acid is perhaps the simplest.

1. Preparation
   a. Wash the skin in soapy water if necessary to remove blood and dirt.
   b. Remove flesh and fat from the skin.
2. Pickling
   a. Prepare 4 litres of pickling solution by mixing 625 ml of vinegar and 450 g of salt in 2.6 litres of water.
   b. Soak skins in pickling solution for 4 to 5 days. Stir at lease twice daily.
   c. Remove skins and scrape off any flesh or fat.
3. Tanning
   a. Prepare a tanning solution in a 4 litre container. Mix together 2.5 to 3 litres of water and 230 g of dry leaves that contain high levels of tannic acid (e.g. tea or oak leaves). Bring to a boil and simmer for about 20 minutes. Allow to cool. Add 625 to 750 ml of vinegar. Add 570 g of salt.
   b. Squeeze out the pickling solution and add skin flesh side down to the tanning solution for 3 days, stirring daily.
   c. Rinse to remove leaves; stretch on a board flesh side out until dry.
   d. Using the hands, work and rub the skin until soft.
   e. Thoroughly wash the skin in mild vinegar solution (1 part vinegar to 16 parts water) to remove excess salt.
   f. Re-stretch and re-dry, working the skin during the final phase of drying until soft and pliable.

Check your learning   Can you:

● Tan a rabbit skin.

XII   GROWING WORMS (VERMICULTURE)

Purpose   Learn to produce worms from rabbit and other manure.

Worms can be grown as a source of high quality protein for rabbits, chickens or pigs. Worms can also reduce the odours and mass of manure from farm animals.

1. Over an impervious surface, prepare the worm bed:
   a. Build a frame 1 m × 1 m and at least 10 cm high.

b. Add enough manure (0.1 cu m) to make a layer 10 cm thick.

c. Break up any clumps in the manure and sprinkle with water to bring the moisture uniformly to about 30% and maintain this level throughout the growing period.

2. Add the worms to the top of the bed.

3. Each week add an additional layer of manure (about 5 cm).

4. After 3 to 4 months harvest the worms by working though the manure and removing the worms. Thereafter worms can be harvested each month.

Check your learning   Can you:

● Produce worms using rabbit or other manure as a feed source.

11

GOAT PRODUCTION PRACTICES

USEFULNESS

Goats are frequent objects of neglect and even prejudice, yet have usefully served mankind with meat, milk, hair, leather and other products including manure.

Because of their small size and affordable nature, goats can be a source of high quality protein for rural families. Children especially can benefit from the better nutrition that goat milk and meat can provide. Goats are small enough so that the carcass can be consumed within a short period of time, reducing the risk of meat spoilage. Compared to larger animals, the lower costs per animal makes the goat more affordable to the small land holder. Furthermore, the death of one of the animals can be less catastrophic.

Goats also serve as a financial reserve to fall back on in case of the failure of cash crops. They serve as a reservoir for food energy and protein produced by crops. Since the goat is so opportunistic and selective in what it can and will eat, it survives when other animals perish. When properly managed, goats can serve an ecological function in the control of brush and other undesirable plants. In Africa, trypanosomiasis resistant goats have been used to clear the low brush which is a favoured habitat of the tsetse fly. The goat is a curious, gregarious and hardy animal requiring little shelter from the weather.

In addition to milk, meat and hides, goats also produce manure that can be a valuable resource of fuel and fertiliser. An 18 kg goat can produce 74 kg of manure dry matter with 1.5% nitrogen, 1.5% phosphate and 3% potash in a year. Although primarily a browser, the goat readily adapts itself to grazing and therefore will fit into a wide range of circumstances. In form and function, the goat is enough like a cow and sheep that information, if tempered with some judgement, can be transposed from one species to the other.

There are some disadvantages to the goat. Goats do have a relatively low individual value and productive capacity in relation to their labour and other cost requirements; their small size makes them susceptible to predators, both animal and human; they are unsuitable as a beast of burden; when concentrated in larger numbers they become susceptible to various diseases and parasites; and because of their rather indiscriminate browsing and grazing habits they can do ecological damage if their feeding is not properly controlled.

210

SELECTING A BREED

There are many breeds of goats (over 60, in fact) each displaying their own peculiar characteristics. Tropical breeds have developed primarily through genetic isolation and natural selection. This has produced an animal capable of survival, but one unable to adequately consume enough feed to support a high level of production. The breeds that thus developed are smaller and have such low yields of milk that, instead of being regarded as dairy breeds, they should be regarded as specialised meat breeds. Angora goats produce mohair that can be used in producing coarse fabrics. Dairy goat breeds have short hair coats, larger body scale and appetite, and highly developed mammary glands. As with most other classes of livestock, the differences between families and individuals within the breed are greater than the differences among breeds of the same type. Emphasis herein will be placed on the higher producing dairy breeds.

Although goats were probably first domesticated in south-western Asia, the breeds that have been developed primarily for milk production originated primarily in Europe. Some exceptions may be the Jamnapari, Beetal and Barbari of India, the Sudanese Nubian of East Africa and the Damascus of the Mediterranean region. The Jamnapari from India has produced 235 kg of milk in 261 days with a maximum yield of 544 kg in 250 days (NDRI, 1976). Goats of the Brazilian Moxoto or Black-Back breed are kept for their skins and meat while being moderate milkers. They can kid 3 times in 2 years with 40% twins.

European breeds that have been crossed with native breeds to improve the genetic capacity to give milk include the following:

Saanen is a white-cream coloured animal, sometimes sensitive to sunlight. It is probably capable of producing the most milk of any breed, but with a low milk fat content. The world lactation record was made by a Saanen doe in Australia with 3084 kg of milk with 3.5% milk fat produced in 305 days (3500 kg in 365 days). Tropical production has been reported as 700 kg in 344 days. This is about half the production that could be expected under good management in temperate zones.

Nubian (Anglo-Nubian) originated in North Africa. It produces less milk but this milk has the highest fat content of any of the breeds. Reported production was 300 kg in 300 days in the tropics.

Alpine, Toggenburg and La Mancha are intermediate between the Saanen and Nubian in their milk and milk fat production. The La Mancha developed in California in 1960 by crossing a Mexican earless mutant buck with established breeds. La Manchas have short to no external ears.

Breeds of goats vary in size from 10 kg body weight and 45 cm withers height for pigmy goats used for meat production to 115 kg and 110 cm withers height in dairy breed females. In America the pigmy goat from

Africa has become popular in zoos, as a back-yard pet and as a meat and milk animal. It is small with a stocky build.

SELECTING AN INDIVIDUAL ANIMAL

Animals can be selected on the basis of type (what they look like), production (what they do) and pedigree (what their ancestors and relatives have done).

Appearance

The parts of a goat are shown in Figure 43 and aspects of conformation in Figure 44.

In selecting the individual milk goat, choose larger animals that show a strong, thrifty appearance. They should have sharpness, angularity and length that is associated with the capacity to convert feed nutrients into milk rather than into body fat; thus avoid a doe that is fat when lactating. The barrel should be long and deep indicating a capacity to consume roughage. One-quarter to one-third of the variation in milk production has been attributed to variations in body size (not body weight). The animal should walk easily and freely so it can forage for feed. The legs should be well apart, joints sound with no swelling. The pasterns, while having some angle, should not allow the dewclaws to touch the ground. The hoof should have depth in the heel and a level sole.

The udder is of special concern. It should be pliable and soft, show capacity and be strongly and closely attached to the body. It should be free of lumps, atrophy or other evidence of mastitis. Teats should be of moderate size for convenient milking (Figure 45).

In the absence of scales for weighing, the body weight of improved breeds of dairy goats can be estimated from the heart girth. Measure the distance around the body behind the forelegs and apply the following conversion values.

Heart Girth		Body Weight	
in	cm	lb	kg
17.5	44.5	20	9.08
20.0	50.8	30	13.64
22.5	57.2	40	18.16
24.0	61.0	50	22.70
25.5	67.8	60	27.24
27.5	69.9	70	31.78
29.0	73.7	80	36.32
30.75	78.1	90	40.86
32.25	81.8	100	45.50
34.75	88.3	125	56.75
37.25	94.6	150	68.10
39.75	101.0	175	79.45
42.25	107.3	200	90.80

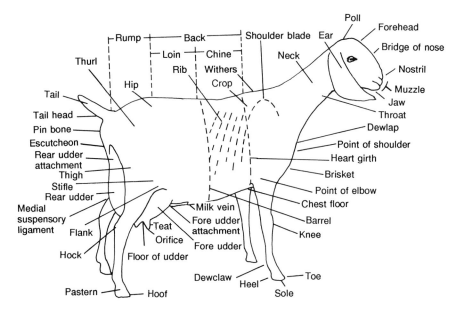

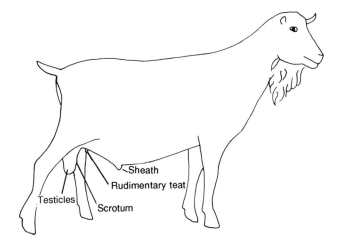

Figure 43 **Body parts of a doe** (above) **and buck goat** (below). *(H. Considine in Extension Goat Handbook, 1984.)*

213

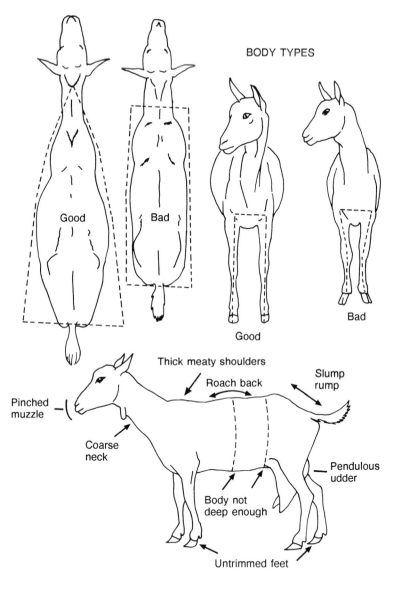

BODY TYPES

Good

Bad

Good

Bad

Thick meaty shoulders

Roach back

Slump rump

Pinched muzzle

Coarse neck

Body not deep enough

Pendulous udder

Untrimmed feet

A DOE OF POOR CONFORMATION

Figure 44   Comparisons of goat types. *(K. Leach in Extension Goat Handbook, 1984.)*

214

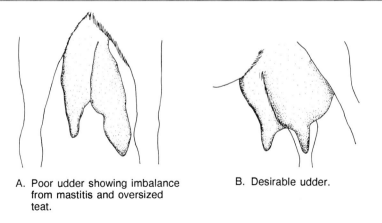

A. Poor udder showing imbalance   B. Desirable udder.
   from mastitis and oversized
   teat.

Figure 45   Types of goat udders.

Production

Actual milk production records, when available for the individual or her relatives, should be relied upon in selection. These records are even more valuable if they had been made under conditions similar to those under which the animals will be expected to perform.

High milk yields are related to high dry matter (DM) intakes by goats. Even with adequate feed available, in the tropics indigenous breeds do not produce as much milk as imported dairy breeds. The imported breeds do not perform as well in the tropics as they do in temperate zones because at higher temperatures the animals will not eat enough feed to maintain high production levels.

Breed	DM Intake as % of Body Weight	Average Milk Yield	Length of Lactation
Tropical breeds in tropics	3.3%	280 kg	199 days
European breeds in tropics	3.6%	570 kg	221 days
European breeds in Europe	4.6%	1476 kg	455 days

In situations where an unfavourable environment can be only slightly improved, crossbreeding might be carried out. Best results come when bucks of the exotic breeds are crossed with indigenous (native) does. This can produce an animal with greater resistance than the exotic breed and at the same time with more productive capacity than the native goats. Results of a trial in India showed:

Indigenous breeds produced   162 kg milk in 167 days.
Crossbreeds produced   375 kg milk in 262 days.
European breeds produced   417 kg milk in 262 days.

215

For optimal production, animals must be genetically blessed with large appetites and the capacity to convert large quantities of feed into milk and other animal products.

Animals with high dry matter intake and high milk yields are more efficient converting feed into milk; the maintenance requirement depends on the body size and remains the same, regardless of production level. The lower the production the higher will be the proportion of the feed required for the overhead (maintenance).

The longer the doe remains lactating the greater the total production will be for any lactation period. Similarly, more frequent milking increases production. In France, does milked only once daily produced 324 kg of milk in a lactation period of 258 days whereas those milked twice daily produced 590 kg in 270 days.

Age

Age has a significant effect on milk production. Yields increase from the first lactation until the second to fifth. After maturity at 4 to 5 years of age, yields decline. For example, if a spring born doe started giving milk at 12 months of age and produced 600 kg, she would be expected to produce 1.33 times as much (798 kg) at maturity. This level would then decline with aging.

The age of a goat can be estimated from the type and condition of the teeth. As with other ruminants, goats have no top incisors, but they do have 8 teeth on the front lower jaw. In animals less than 1 year of age these teeth are small and sharp. Two large permanent teeth will replace the centre pair at about 1 year. Two more teeth, one on each side of the 1 year teeth, appear at about the second year. There are 6 permanent teeth in the 3 to 4 year old. By the fourth to fifth year the goat has all 8 permanent teeth in front. After this, the age is estimated by the amount of wear and cup formation on the teeth. As the animal ages, the gums shrink away, the teeth spread apart and loosen with some dropping out, thus reducing the ability to browse.

Pedigree

The pedigree, or record of ancestry, can be used as a basis for selecting individual goats. A pedigree should have production and other information about an individual, as well as the ancestors. This type of information is helpful in estimating the genetic value of an animal. Performance data of only parents, grandparents and siblings are of significant value.

BREEDING

In many areas, indigenous goats have developed with inadequate, unpredictable feed supplies, high levels of disease and parasite infection, harsh climates and other unfavourable environmental conditions. Thus, the animals have evolved with a capability to survive. This resulted in a genetic makeup that leads to slow growth and maturity rates, and only enough milk production to assure the survival of offspring. With the improvement of environmental conditions, some increase in growth and milk yield can be expected. However, this increase is limited by the genetic capability. Genes for high production can best be introduced by crossing indigenous does with bucks that possess this capacity.

In Kenya it was reported that the first cross between Toggenburg bucks on Galla and East African breeds was about average between the parents, or 3.2 to 3.6 kg milk per day at peak lactation. Hybrid vigour was also observed.

In goats short generation intervals, often less than 2 years, together with frequent multiple births make possible a rapid genetic progress through selection. Multiple births, however, provide increased selection and production potential only if the offspring live and reproduce themselves. Under limited feed conditions, multiple births can be a disadvantage because they increase the stress on the dam, reducing her productive life.

Any imported exotic breeding animal should be a young, rapidly growing one showing desirable physical characteristics. Genetic potential should be verified with milk production records of the dam and proof of the sire's ability to transmit production ability to his offspring. The imported animal should also be free of disease and meet any health standards set by the importing country. Only horned bucks should be used in a breeding programme because of the association of hermaphroditism with polled (hornless) bucks.

If technology is sufficiently advanced, frozen semen can be imported from bucks proven to be capable of transmitting high production to their offspring.

REPRODUCTION

The structures of the reproductive system of the female and male goat are depicted in Figure 5 and Figure 8, pp. 17 and 22.

The level of reproductive performance depends on both genetic and environmental factors, and how they interact. The heritability of reproductive effectiveness is only about 10%. This means that 90% was due to management, climate, feed and other environmental factors. In fact, the reproductive performance of an animal is an excellent measure of the level of its adaption to the tropical climate. Heat and poor nutrition have a negative effect on reproduction, resulting in prolonged kidding intervals and fewer and smaller kids.

Does

Well-fed does usually enter puberty the autumn following their birth in the spring. This first oestrus can come as early as 4 to 5 months of age. However, most goats are not bred until the second breeding season 12 months later. Thus, although doelings can be mated when 7 to 10 months of age, they should be 1 year or more old. However, onset of puberty is more a function of body weight than age. Puberty occurs when doelings have reached 65% to 75% of the weight of the mature does in the herd. Therefore, because of the seasonal nature of their reproductive behaviour, does should be grown as rapidly as possible, so that they could be mated before 1 year of age. Faster growth with earlier kidding permits does to come into production sooner. This not only provides a greater population turnover and thus allows more rapid genetic progress, but it increases the lifetime milk productivity.

Most tropical breeds of goats come into oestrus throughout the year. However, temperate climate breeds of goats are seasonal breeders, becoming sexually active as the days grow shorter. The breeder can reduce this tendency for cyclical breeding by selecting those animals that come into heat early and/or late in the breeding season. This cyclical breeding effect declines as animals are moved toward the equator. Two kiddings per doe per year are theoretically possible were the animals not seasonal breeders and environmental conditions optimal. Tropical indigenous breeds are less inclined to be seasonal breeders and 3 kiddings every 2 years are common.

Heat (oestrus) in the doe is characterised by uneasiness, bleating, riding other animals, standing for riding by others, vigorously shaking an erect tail, frequent urination, and displaying a red swollen vulva with the presence of some mucous. The presence of an odouriferous mature buck tends to bring does into oestrus at an earlier date.

A mature doe is in heat for 18 to 36 hours every 18 to 22 days during the breeding season. The presence of a buck hastens the onset of oestrus. The oestral cycle can be interrupted if does are moved to unfamiliar surroundings during the breeding cycle. Mating should occur toward the end of oestrus (e.g. at least 12 hours after oestrus is first observed). Copulation occurs rapidly, with intromission and ejaculation requiring only a few seconds. Many does, even if pregnant, will accept the buck.

Bucks

Buck kids should be separated from doelings at 2 to 4 months of age to prevent premature mating. A buck, if well fed, may reach puberty at 3 months and achieve adequate size and maturity for light service at 6 to 10 months. Healthy bucks have exceptionally high libido and infertility is common only in polled bucks (which in some breeds are usually genetic intersexes).

During the rutting (mating) season, the buck has some rather disgusting behaviours of urinating and ejaculating on his chin and underside which increases the odour emitted by his musk glands. Clipping the hair on the head, neck and belly tends to reduce this unpleasant odour. Because of this strong odour, bucks should be kept 50 to 75 m away from the dwelling and the does. If does are in physical contact with the buck these odours will be transmitted into the milk. They can be reduced somewhat by removing the scent glands that are located medially and caudally to each horn. This can be done easily at the time of disbudding (dehorning) the young buck if the hot iron is used. The odour of these glands and that of mature bucks will penetrate the skin and remain for several days. Consequently, gloves are sometimes worn when working with these animals.

Intersex

Hermaphroditism is a frequent source of infertility in dairy goats. These intersexes are more often genetic females having XX sex chromosomes. As such, the vulva is normal in size, but the vagina is underdeveloped and the clitoris is enlarged. Both hermaphroditism and congenital hypoplasia are associated with the polled (hornless) condition. These defects are more likely to occur when both parents are polled. The frequency of the polled gene is higher in some breeds such as the Saanen, while being rare in Angoras. Hornlessness is dominant over the horned condition, which facilitates selection against this trait; if a dominant gene is present, it can more easily be detected.

Hornlessness is seen as an advantage because there is no need manually to disbud the goats. However, this advantage is offset by the losses due to hermaphroditism.

Artificial Insemination

Artificial insemination has proven successful in some of the more developed areas. Goat semen has been successfully diluted and frozen. The advantages of artificial insemination include:

1. Wider use of genetically superior sires.
2. Keeping a buck becomes unnecessary, which reduces feed costs, special fences and housing, and odour.
3. Reducing likelihood of disease transmission when animals are moved from one farm to another for mating.

Gestation

The average gestation period is 145 to 155 days (average 150). The number of offspring is determined both by genetics and environment. Feeding so that the doe is gaining weight at breeding (flushing) will increase the incidence of multiple births. On the other hand, starving early pregnant does for 2 days can cause death and re-absorption of a high percentage of embryos.

Twins are common but occasionally triplets, or even quadruplets, are born to well-nourished does. Young, smaller does produce fewer kids and experience higher abortion rates and kid mortality. Does born as singles usually kid 1 month earlier than twins. Larger does have a higher percentage of kids raised and weight of kid weaned. Undernutrition reduces both the length of gestation and the size of the foetus.

Kidding

About 140 days after breeding the doe should be making preparation to kid. The udder should become distended. With the first kidding there is usually considerable udder oedema developed, especially in higher producing does. If the udder becomes painful, the doe might be milked, but this colostrum should be saved for the kid. The vulva will become enlarged and the ligaments around the tail head will relax. The doe should be confined in a clean, dry pen and left alone.

Symptoms of kidding are uneasiness, bleating and pawing. First, the cervical plug of mucous is voided. This together with the fluids released when the placental sac breaks lubricates the vagina. Thereafter, the kids are usually born within an hour. If parturition has not occurred after 1.5 to 2 hours after the water has broken, or if there is no outward movement of the kid in 5 or 6 labour strains, assistance is probably needed. The location of the pregnant uterine horn in relation to the rumen and pelvis is shown in Figure 6.

To assist, secure the doe's head by tying or placing in a stanchion. Having trimmed fingernails, wash both hands and the doe's perineal area with soap and water and rinse with a sanitising solution. If the lubricated hand can be inserted through the cervix, the doe should kid. The proper presentation of the foetus at birth is with the head between the two front legs (Figure 46). If the foetus is not in a normal position some manipulation to get it into the normal position will be necessary. The abnormal foetal positions that are usually encountered include head bent backward, one or both front legs back, or rump first (breech) with legs bent forward over the belly. These require outside assistance to be re-positioned. If both hind feet are coming out first a breech birth can be successful. It is necessary to be able to recognise the difference between a knee and a hock. Any

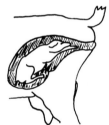
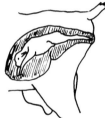

Normal position

Breech, rear
feet first

One leg forward,
one leg backward

Both front
legs backward

Head turned

Breech, rump first

Figure 46   Normal and abnormal foetal presentations. Delivery usually occurs with the kid in the normal position or breech with both back legs coming out first. All other presentations require assistance.

manipulation should be done when the uterus is not contracting. With proper presentation let nature take its course. Sometimes with first kidders or unusually large kids, assistance in passage may be necessary; in this case the ends of a sanitised obstetric chain or 1 cm wide smooth nylon or cotton cord are tied to each foot. Pressure is then applied to the foetus by pulling backward and downward from the doe with each straining effort she makes. Then hold the tension, ratchet-like, until the kid has been delivered.

The weight of the kid at birth is positively correlated with the size of the dam but negatively correlated with the litter size. Although the individual kids weigh less, the total weight of newborn increases with their number.

The afterbirth will normally be passed within 4 hours. If this has not occurred after 6 hours, some gentle assistance or professional help from a veterinary surgeon may be in order to prevent uterine infection and infertility.

A good practice as soon as the kid is born is to dip its navel in a dilute iodine solution to prevent passage of pathogens into the abdominal cavity, bloodstream and other parts of the body.

Mortality in goats is highest from birth to weaning. This suggests the importance of management during this time. Fertility as well as milk production improve with age up to full maturity, after which they decline.

Kid Rearing

A low mortality rate and producing a doe weighing 45 to 50 kg at kidding should be the goal of a kid management programme. It is sometimes desirable for the number of milking animals to be kept constant. If dairy goats are to remain for 4 lactations, one-quarter of the animals must be replaced each year. The number of kids required to maintain the flock numbers is increased by death and other losses. Kid mortality has a direct effect on the selective pressure that can be applied for genetic improvement.

Does give birth about 149 to 150 days after being bred. A few days before expected kidding the doe should be isolated in a clean, dry, bedded area provided with water and good hay.

In cold weather, the newborn should be protected from chilling. They can be dried with a clean cloth and placed near the doe where she can lick them.

It is important that the kids consume the first milk (colostrum) that is produced by the dam after kidding. For the first 2 or 3 days kids should receive a daily total of up to 10% of their body weight in colostrum in 4 or 5 increments. This first milk should be given whether the kids are left with their dams or separated at birth. Colostrum provides passive immunity as well as being a rich source of protein, minerals and vitamins. This is nature's way of imparting passive immunity in the young by transmitting antibodies from the dam to the newborn. Antibodies are modified gamma-globulins, a special kind of protein. The antibodies against the diseases to which the dam is immune are produced in the pre-parturient udder and for the first few hours of life these are not digested by the kid but can be absorbed intact from the intestine into the bloodstream.

If the kid has not suckled within 2 hours of birth, check the doe's udder and teats to make sure there are no abnormalities. A retarded kid can be taught to suck by placing the end of the dam's teat into its mouth and squeezing milk from the teat.

When first born, the kid is essentially a simple- stomached animal having much the same nutritional requirements as humans. The kids should be allowed to nurse for 1 to 3 days, then taught to drink from a pan. Milk should be fed at the same temperature, preferably 39°C to 40°C. *Do not overfeed*; 10% of a kid's body weight daily should be adequate. Goat milk or a milk replacer can be fed. The liquid replacer should contain 14% solids from a preparation containing 20% animal fat and 24% protein and less than 1% crude fibre in the dry matter. For the first month feed thrice daily and twice daily thereafter. Provide the kid with some good quality roughage and, if available, a concentrate.

Traditionally, kids are weaned at 3 months of age. However, with appropriate management they can be weaned at 9 kg body weight, 8 weeks of age or at the time they are consuming at least 30 g of solid feed daily, whichever comes first. Solid feed consumption is encouraged by restricted milk feeding. Solid feed is required to establish a microflora and microfauna and to make the structural changes in the stomach itself to be characteristic of the mature ruminant. Since the rumen is not yet fully functional, some grain or other concentrate should be fed. Rapid growth to 38 to 40 kg for first breeding is dependent upon an adequate intake of digestible nutrients.

Buck kids that are not to be retained for breeding purposes should be castrated; the earlier in life (7 to 14 days of age) the less severe is the operation. Disbudding (dehorning) and removal of scent glands can occur at the same time.

FEEDING

To illustrate the importance of the forestomach to the goat, of all the nutrients digested, 58% of the dry matter, 93% of the crude fibre and 81% of the soluble carbohydrate are absorbed from the rumen–reticulum–omasum complex. The balance of the digestible fraction is absorbed from the intestine and caecum. Of the gross energy consumed by the lactating doe, typically, 30% is lost in the faeces, 5% in urine, 20% as heat and 5% in combustible gas; 20% of the feed energy is used for maintenance and the remaining 20% is available for production (Figure 47).

Nutrient Requirements

As to the specific nutrient needs of the dairy goat, maintenance requirements depend upon animal size, activity level and adverse environmental factors. Maintenance accounts for about 50% of the total feed energy required for meat production. Goats with a greater milk production potential will have a higher rate of metabolism and thus will have a higher

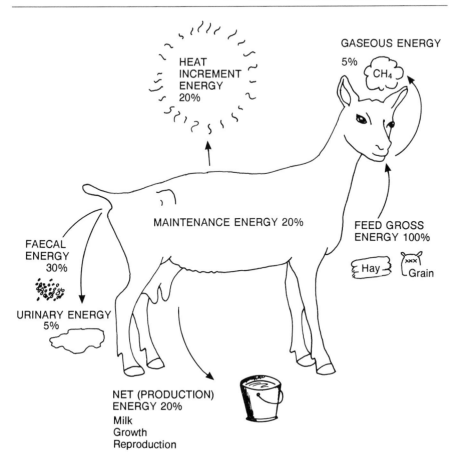

GASEOUS ENERGY
5%

HEAT
INCREMENT
ENERGY
20%

MAINTENANCE ENERGY 20%

FEED GROSS
ENERGY 100%

FAECAL
ENERGY
30%

URINARY ENERGY
5%

NET (PRODUCTION)
ENERGY 20%
Milk
Growth
Reproduction

Figure 47   Utilisation of feed energy by the goat.

maintenance requirement per unit of body weight. The requirements for production will vary with the growth rate, pregnancy, amount of milk produced and the milk fat content. Feed energy is converted into milk energy more efficiently than it can be converted into body fat energy of the non-lactating animal.

The requirements and digestive physiology of ruminant goats, sheep and cattle have been found to be so similar that information learned on one species can generally be transposed to another. Tables have been prepared for estimating the nutrient requirements under various conditions. Considering only body size and milk production, daily energy requirements for maintenance are 123.63 kcal digestible energy per kg metabolic weight (body weight raised to the 0.75th power), and 0.032 g total protein per kcal digestible energy. Additional daily requirements to rebuild depleted body stores during late pregnancy would be 1.74 Mcal DE and 82 g protein.

223

Additional daily requirements for each gram of growth gain in young does would be 8.8 kcal DE and 0.28 g protein. The digestible energy requirement per kg of 4% fat corrected milk is 1460 kcal. For each 0.5% change in milk fat content from 4%, an addition or subtraction of about 115 kcal digestible energy applies for each kg milk produced. These values have been converted to practical tabular form in Table 12.

Table 12 Daily Nutrient Requirements of Lactating Goats

Variable	DE kcal	TDN g	ENE Mcal	Crude Protein g	Ca g	P g	Carotene* mg/kg
For maintenance plus low activity level (125% of maintenance)							
Body weight (kg)							
10	870	199	0.40	27	1	0.7	1.25
30	1990	452	0.92	62	2	1.4	3.0
50	2920	662	1.34	91	4	2.8	4.5
70	3760	852	1.75	118	5	3.5	6.5
90	4540	1030	2.09	142	6	4.2	7.0
Additional requirements for milk production per kg at different fat %							
% fat							
3	1230	337	0.60	64	2	1.4	9.5
4	1460	346	0.70	72	3	2.1	9.5
5	1680	356	0.72	82	3	2.1	9.5
Additional requirements for late pregnancy (for all goat sizes)							
	1740	397	0.70	82	3	1.4	2.75
Additional requirements for growth							
Daily gain (g)							
50	440	100	0.20	14	1	0.7	0.75

* 1 mg carotene = 400 IU vitamin A.
Note that the 'additional requirements' are the amounts to be supplied in addition to maintenance.

Thus, nutrients are needed by body cells for growth and maintenance, by the mammary glands for milk synthesis, by the foetus for reproduction and by fat tissues for body energy reserves. It has been estimated that with each increase in the milk fat content of 1 percentage unit the requirement for protein and energy for milk production will increase 12% and 16%, respectively.

Intake Level

Factors of practical importance that influence intake include body weight, change in body weight, energy concentration in the ration, the bulk and density of the ration and environmental temperature.

Voluntary consumption is influenced by genetic capacity and several other factors such as access to the more preferred species of plants, maturity and digestibility of forage, and a comfortable temperature and humidity. Spoiled mouldy feed is not only unpalatable but may actually be harmful if eaten.

The growth and production levels are frequently limited by the inability of the animal to consume enough dry matter. Some mature animals have been reported to consume more than 5% of their body weight; however, many, especially kids, cannot consume more than 3%.

It is essentially impossible for a doe in high milk production in early lactation to eat enough to meet her demands for both body maintenance and milk production. She must draw upon her body reserves to make up the difference, losing weight in the process. This weight is recovered during late lactation and the dry period. This points up the need for does to be in a thrifty, but not fat, condition at kidding. A note of caution: pushing heavy concentrate feeding in the first 14 days postpartum can cause metabolic problems and illness. The association of the typical milk production curve, dry matter intake and body weight changes during the first 30 weeks of lactation is illustrated in Figure 48.

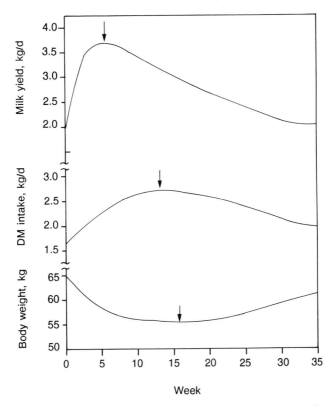

Figure 48 *Changes in lactation curve, dry matter intake and body weight of dairy goats during lactation.*

Roughages

Peak milk production occurs at 6 to 8 weeks postpartum; however, peak feed intake does not occur until 10 to 14 weeks postpartum. Because they are unable to consume enough feed to meet the need, heavy producing does must draw on body reserves which leads to losses in body weight. The protein and energy concentration of the feed that is eaten early in lactation should be higher than at other times. High concentrate diets might be fed in an effort to meet this increased need. However, the capacity of the high producing goat to consume concentrates to meet the high energy demands is limited. As grains make up an increasing proportion of the diet, the goats will first produce more milk with a lower fat content, then with increased energy concentration in the diet, the milk production drops and the energy intake over maintenance is deposited as body fat. As illustrated in Figure 49, the maximum efficiency with which feed energy is converted to milk energy is when grain makes up 40% to 60% of the dry matter intake. However, for fattening the efficiency for converting feed into body gain continues to increase as roughage is replaced by energy feeds. The practical limit is reached when roughages make up only 15% to 20% of the total diet.

Dairy goats should not carry excess flesh; fat goats are more susceptible to ketosis and similar problems.

Between 50% and 75% of the goat's energy is absorbed from the rumen in the form of the volatile fatty acids (acetic, propionic and butyric).

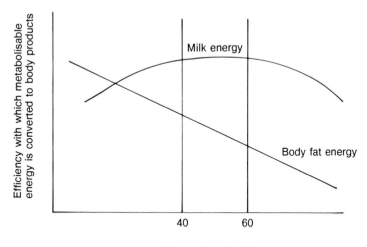

Figure 49   Efficiency with which feed energy above maintenance is converted to energy in milk and body gain. Increasing dietary grain until it makes up to about 50% of the diet increases the efficiency with which milk is produced. Beyond this point the efficiency of milk production declines but the efficiency for body growth and fattening continues to increase.

226

As with other herbivores, goats have a fibrosity requirement that can be met by the inclusion of about 17% crude fibre and some long, unchopped forage in the diet. They can flourish and maintain mediocre production on good quality roughage alone, but higher production requires the feeding of some concentrate.

Compared to grass, browse (leaves and twigs of trees and shrubs) and forbs (non-grass weeds or broadleaf herbs having no woody stalk) generally contain higher levels of crude protein and phosphorus during the growing season. However, the value of many palatable browse plants is often limited because of one or more inhibitors that might be present. These inhibiting factors make unavailable certain nutrients, or they may actually be toxic. In some range shrubs essential oils might inhibit growth of rumen bacteria, and tannins depress digestion. Lignification of woody twigs and leaves physically binds or encapsulates the nutrients. The effects of these negative factors are minimised by selective browsing when the goat has this opportunity.

The goat ration should generally be based on roughage because of its wider availability and usually lower cost than concentrates. Improved pasture of legume–grass mixtures over which the animals are rotated provide a choice feed source. The animals are to be exposed only to the amount of fresh plants that they will clean up that day. This rotation not only allows the plants to restore themselves but also reduces the likelihood of infestation from internal parasites by allowing the sun to dry out any worm eggs. In drought-prone areas fast growing trees can be planted to provide browsing although it is generally difficult for high producing goats to consume enough dry matter from browse alone. By standing on their hind legs, goats can browse up to 2 m on trees and hanging vines.

Confining goats to fenced, improved pastures is often not feasible because of the expense involved in building a goat-proof fence. The goat's desire to browse for a more varied diet than grass alone further complicates the practice.

Tethering can be used where there is no danger of the tethering chain becoming entangled in tall, growing plants. Regardless of the management system, shade and water must be available to the goat as well as protection from rain in cold weather. Animals do better when water is constantly available than when it is offered only twice daily. A 10% reduction in milk production can be expected when goats are watered only once daily. It is interesting to note, however, that non-lactating goats can go for long periods of time without water. Furthermore, they have the remarkable ability to drink up to 45% of their total body weight in 2 minutes without noticeable ill effects!

A constant or intermittent 'zero grazing' (dry lot feeding in which all the feed is brought to the animal) management system offers many advantages for small and family herds in heavily populated and/or cultivated areas. Such things as industrial or crop wastes or vegetable peelings are often available for feeding. Where there is no opportunity for grazing, nor labour for herding, this no-grazing system provides more convenient control of feeding and mating as well as close supervision of kidding. Higher producing exotic breeds and their crosses frequently do better under this more intensive system.

Zero grazing, then, has the advantages of:

1. Conservation of animal energy which would otherwise be wasted in roaming in search of often poor quality feed.
2. Reduced erosion from trail making.
3. Salvage of manure for bio-gas formation, fertiliser or worm production.
4. Better animal management with individual feeding, care and predator protection.

Constraints of this system are that the feed must be harvested, stored and brought to the animal. This requires additional labour and management capabilities.

The rule of thumb states that roughage dry matter should be fed at 2% of the body weight each day and 1 kg of grain be fed for each 2 to 4 kg of milk, depending on the roughage quality. Due to differences in moisture content, 4 kg green chop or 3 kg silage is equivalent in dry matter to 1 kg dry hay.

Concentrates

Concentrates should be fed to supply any nutrients lacking in the roughages. Energy is the factor most likely to be deficient. It can be supplied by grain or other high energy feedstuffs.

Good quality legume roughage with some grain will generally provide all of the protein and calcium a goat will need without special supplementation. However, if grass roughages are fed, the grain should contain 16% to 18% protein and 1% calcium. All rations should contain 0.5% salt. In addition, a mineral box, protected from the elements, should offer the animals a free-choice calcium–phosphorus supplement (i.e. steamed bone meal) and salt (preferably trace-mineralised).

Although goats are relatively sensitive to urea feeding, if care is exercised to completely mix the feed, urea can be included as a partial protein replacement providing it supplies no more than one-fourth of the total crude protein in roughage type diets and not more than one-third of the protein in the concentrate portion of the diet. Then, the animals must be adapted by gradually increasing the amount of urea over a 3-week period.

Does with their first pregnancy and lactation will require extra feed to support their continued growth. Depending on how well grown they are at kidding, first lactation does should be fed up to 120% of their maintenance requirement. Some extra feed (1700 kcal DE and 80 g protein) should be provided to pregnant does to offset the demand of unborn kids that put on 70% of their birth weight during the last 6 weeks of pregnancy. At kidding, does should be in good flesh, but not fat.

The importance of adequate feeding was demonstrated by a feeding trial with first lactation Barbari does. Those receiving 100% of their energy requirement and 125% of the protein requirement produced 101 litres in

190 days. Those receiving only 75% of the requirement of energy and protein produced only 27.8 litres in 112 days of lactation.

Kids should have access to hay and grain as early as 1 week of age. Creep feeding or separate housing from adults is recommended.

An alert and thrifty animal with a sleek hair coat is the best indicator of a successful feeding programme.

MANAGEMENT

Characteristics

Goats are individualistic animals and, while being gregarious, they do not have a well-developed herding (flocking) instinct as do sheep. They do not herd well; they turn to face approaching persons or dogs or perceived aggressors. Goats are curious, friendly and responsive to kindness and care. This, together with their intelligence, make them good pets to which people may become attached.

Housing

Housing should keep goats dry, be well ventilated but free from draughts, be light, well drained and easily cleaned. In temperate climates housing needs are minimal. Protection from the sun, rain and, in some areas the wind, is all that is needed.

Neighbouring dogs and other predators are a constant danger for goats, especially the kids. Goats, being the curious, active animals that they are, need freedom of movement and elevated objects on which to run and jump or to push around. Bucks require exercise to retain their fertility. This active nature is a disadvantage in that it complicates their confinement. A fence of woven wire or other material 120 to 140 cm high will hold most goats (bucks may need another 15 to 30 cm height). Gates need to be secured in such a manner that the curious goats cannot open them.

Less desirable than a fence is a tether chain or rope equipped with a swivel to prevent twisting and knotting as the goat moves about. Children can herd the goats as they graze and browse. Due to the eating preferences of goats, they should be kept away from valuable plants. They will literally eat down shrubs and low trees.

A feed manger should be provided with a floor about 30 to 45 cm off the ground and vertical openings just wide enough for the goat to insert its head. Goats are rather fastidious eaters, generally refusing to pick up hay that has fallen on the ground, or that which is spoiled or mouldy. They also prefer eating with their heads elevated.

Pens that can be moved at intervals are desirable for kids.

Milking

Goats are creatures of habit, so much hassle can be avoided by properly training the animal according to the owner's desires. Feeding, milking and other routine factors affecting the animal should be firmly but kindly performed in the same manner on a regular basis.

Goats should be milked twice daily at 12 hour intervals. When in heavy production, milking 3 times will increase production by about 15% with an additional 5% increase with 4-time daily milking.

A practical total milk production of goats in tropical climates might be 150 to 175 kg per lactation. Short lactations of less than 300 days are characteristic.

Since goats are small animals, milking is facilitated by a milking platform such as a bench 0.7 m high, 1 m long and 0.55 m wide provided with a slatted ramp at the rear. A seat for the milker on one side, a railing along the other side, and in front, a stanchion for restricting movement and a feed box for entertaining the doe are all desirable.

All the milk the goat will give at each milking is stored in the gland at milking time, but it cannot be removed without the assistance of the hormone oxytocin to contract the smooth muscle fibres around the alveoli, the process called 'let down'. Oxytocin release is induced by stimuli associated with nursing such as providing an atmosphere free of distractions and massaging (washing) the udder and teats with warm water. Draw a stream or two from each teat across the bottom of the container to check for clotting, blood or other evidences of mastitis. Oxytocin remains in the blood only 5 or 6 minutes, suggesting the need for quick milking. If properly managed, being milked is a pleasant experience for the doe.

Some persistent milkers will continue producing almost indefinitely, but all does should have 4 to 6 weeks rest between lactations to allow for restoration of body reserves and rebuilding the milk secreting tissues in the udder.

As soon as milk is drawn, it should be strained and cooled to about 4.5°C for storage.

Milk is an excellent food not only for humans but for microorganisms as well. Consequently care should be taken to keep it cool, free from dirt, insects, etc. Equipment should be designed with smooth seams and of impervious material (stainless steel preferred) for easy cleaning. After rinsing the milk away with water, equipment can be cleaned with a household detergent and sanitised by scalding or rinsing with a 300 ppm chlorine solution and allowed to drain. After cleaning, the milk equipment should be stored in a dry place free from dust and flies. Milk can be consumed in the fluid form or processed into yogurt, cheese, butter or other products. Pasteurisation of the milk (heat to 54.5°C for 30 minutes or 73.9°C for 20 seconds) will destroy any pathogens present and extend the shelf-life.

If goats are maintained on soft earth their hooves might need regular trimming in as much as sore feet drastically reduce feed intake and milk production. Sharp pruning shears, a farrier's knife and/or jack-knife are effective tools, especially if the hooves have been softened by walking on a wet surface.

Animals should be identified; tattooing is preferred because ear tags are prone to be torn out.

DISEASES

Diseases are conditions that impair normal functioning. After nutrition, diseases are generally the most limiting factor to successful management. Furthermore, they are usually problems of the entire herd or flock rather than a problem of individual animals. Disease may be due to a variety of causes (aetiology) which will influence the symptoms, severity and any type of treatment necessary for their cure. There are numerous diseases brought about by factors both inside and outside the animal's body. It is possible here to list only a few that seem most likely to be encountered. Special emphasis should be placed on those conditions that are commonly experienced in the area.

Internal Parasites

The most economically important parasites within the body inhabit the gastro-intestinal tract, liver and lungs, although there are those that invade other body tissues and the bloodstream. Those in the gut frequently come to equilibrium, doing little damage, being kept in check by the natural resistance of the host. The parasite, however, gains the upper hand when animal resistance is lowered by poor nutrition, severe climatic conditions or other diseases, or when there is a heavy re-infestation of the parasite such as when the animals are grazing for long periods on the same pasture. Warm weather and wet pastures are conducive to parasitic growth. While it is usually impossible to clear the soil of parasites, sunlight and dryness together with pasture rotation and tillage reduce their numbers. These circumstances, then, make treating the animals and making management changes the most feasible means of control. Since parasitism is a herd problem, treating individual animals (i.e. only those obviously affected) is of limited value. Microscopic examination of faecal samples can indicate the kind of internal parasites and their concentration.

There are many different species and varieties of internal parasites. They are seldom a problem with a few goats browsing over vast areas of dry, sun-baked terrain. However, animals repeatedly grazing grass in wet, warm areas or those kept in close confinement with many other animals become highly susceptible to internal parasites.

The symptoms of infestation vary with the parasite and the particular tissue that the parasite is programmed to invade. Whenever internal parasites become a problem, their life cycle must be understood and management measures taken to keep the numbers of worms under control. This usually means using an anthelmintic to rid the goats of adult worms and ploughing and rotating the pasture to destroy the eggs and reduce the level of re-infestation. A systemic treatment that is often effective against both internal and external parasites is a prescribed dose injection of Ivermectin.

The following internal parasites are frequently encountered:

Roundworms (ascarid and strongyloid nematodes) are among the most important internal parasites. Some large strongyloides are blood suckers;

smaller ascarides are not. Most roundworms have a direct life history, i.e. they do not require an intermediary host. How worms re-infest the host depends on the type of worm. Large female roundworms lay eggs that are voided in the faeces to be ingested by the host. They hatch in the intestine, migrate to and grow in the liver, move to the air spaces in the lungs where they are coughed up and swallowed to repeat the cycle. Other mature worms such as hookworms in the stomach or intestine lay eggs which are voided in the faeces. On the ground they hatch after a few days or years, depending on the temperature and moisture. The larvae require 2 to 10 days to pass through a maturation stage; they then crawl up stalks of grass to be eaten by a new host. After maturing for 2 to 6 weeks in the animal tissues or digestive tract, the cycle is repeated as the parasite begins laying eggs.

Usual signs of infestation include a general unthriftiness, a rough hair coat and a run-down condition. The animals lose weight and have a poor appetite. Although other species usually have diarrhoea, goats seldom do even though heavily infested. Anaemia may also be evident from paleness of lips, tongue and mucous membranes of the eyes. A chronic cough and/or a swelling beneath the lower jaw may also be evident.

Lungworms (*Dictyocaulus*) are especially prevalent in cool, wet areas and can be a severe disease problem. They cause coughing, pneumonia and weight loss. Levamisole (8 mg/kg body weight) is an effective treatment.

There are many anthelmintics that are safe for goats; thiabendazole and levamisole are two. Routine worming (2 doses 2 weeks apart) at periodic intervals to control stomach and intestinal worms is recommended where worm infestations are high. Animals on permanent pasture are especially susceptible.

Coccidia (*Eimeria*) are microscopic protozoa (single-celled animals) that inhabit the intestinal tract. Coccidia in goats, rabbits and chickens are not cross-infective; that is, goats do not get coccidia from other species (except sheep), and vice versa.

Coccidia have highly complicated life cycles by which they reproduce and invade the intestinal mucosa of the host. Oocysts are passed in the faeces. If the faeces are examined under a microscope, the presence of large numbers of oocysts is indicative of the disease. These oocysts can easily contaminate feed and water. Infestation is marked by a lack of appetite, chronic diarrhoea (which is often bloody), and anaemia, all of which can make the animal lose weight, become dehydrated and die. Secondary infections including pneumonia are common complications.

Coccidiosis is usually limited to kids under 4 months of age although older animals that have not previously been exposed will display symptoms 2 to 4 weeks after exposure. All adults can carry and spread coccidia without showing any symptoms themselves. For this reason, to prevent their infestation, young goats should be kept separate from older ones. Treat all adults to prevent them shedding coccidia. Administer 110 mg sulphadimethoxine per kg body weight the first day and 55 mg per kg body weight on the second, third and fourth days.

Control of coccidia depends on sanitation and management to keep kids in good health and to prevent buildup of coccidia in the soil. Sulphonamides and nitrofurazones are excellent coccidiostats to be best applied before stress factors are encountered; they retard the growth but do not kill the parasite. This allows the animal to develop its own resistance. These coccidiostats can be administered in the feed or water. It is unrealistic to try to eradicate all coccidia from dairy goats kept in confinement.

External parasites

Lice spend their entire life cycle on the goat, whether they suck blood or live off hair fibres and skin debris. Animals can be dusted with insecticides but complete spraying or dipping is more effective. Effective insecticides include coumaphos, malathion and ciodrin.

Caution: insecticides not only kill pests, but also can do damage to both man and his animals. Always carefully follow the manufacturer's directions. Provide safe insecticide storage away from children. Do not contaminate either the milk utensils or the milk with insecticides.

Ticks of many species are widely distributed over the world but occur principally in tropical and subtropical countries. Not only do they sap their host's blood, but they injure their host by the toxic effect of their saliva and by the transmission of diseases to both man and animals. They puncture the skin allowing the entrance of pathogens and other parasites such as screwworm, and at the same time they damage the hide for leather.

Mature ticks can be manually removed if care is exercised not to leave broken-off mouthpieces embedded in the animal. There are many effective insecticides including pyrethrins, rotenone, lindane and toxaphene. Follow the manufacturer's directions in all cases.

Mites that belong to several species cause mange, a contagious skin disease. Mites are small insects less than 2 mm in diameter; a microscope is helpful in diagnosis. The most satisfactory method of control is dipping or spraying the animals. Insecticides such as rotenone and toxaphene are effective. Or, some effective preparations can be mixed at home. A lime–sulphur dip is prepared by mixing 5.44 kg of unslaked lime (or 7.25 kg of commercial hydrated lime; not air-slaked lime) and 10.89 kg of sulphur flour to 380 litres of water. Ivermectin is also effective.

Psoroptic mange (scabies) is caused by mites that bite the skin but do not burrow.

Sarcoptic mange is caused by burrowing mites. Their burrowing and feeding habits cause intense itching, scratching and dermatitis and loss of hair.

Chorioptic species of mites infest goats, some forming lesions primarily on the feet and legs; others affect the ears.

Demodectic mange is caused by a spindle-shaped mite that lives deep in the skin where they give rise to soft oval nodules on the neck, shoulder

233

or flanks. These lesions may be 0.3 to 2.5 cm in diameter, usually with a smooth surface. As soon as these nodules appear they should be carefully cut in a cross pattern with a sharp knife. The cheesy, mite-containing material should be expressed and the cavity painted with formalin or iodine until healed. There is no other satisfactory treatment. Infection is thought to occur when the kid is nursing, but symptoms rarely occur before the animal is 1 year of age.

Ringworm is caused by a fungus transmitted by contact. It produces a thickened, grey, scaly skin with thinned hair coat but no itching (unless accompanied by lice). Ringworm can be treated by applying every other day a mixture of equal parts of tincture of iodine and glycerine.

Infectious Diseases

Mastitis is an inflammation of the udder due to a variety of factors. Predisposing factors include any stress that will reduce the natural resistance, injury to the udder or infusion of highly pathogenic organisms. Infective agents may be bacteria, mycoplasma or yeasts. The infection might be localised in the udder or it might affect the entire system, in which case high temperatures and death can result. The udder becomes sore, hard and swollen. The secretions might be watery with tinges of blood or might contain cheesy clots. Acute forms of mastitis might be treated by making the animal comfortable, milking out the infected udder every few hours and treating with an antibiotic either by infusing it into the affected half of the udder or by an intramuscular injection for a systemic effect. The teat end and opening must be carefully cleaned and disinfected before any object is inserted. As with the use of any drug, the manufacturer's directions should be carefully followed. Cold water or ice packs help reduce swelling and inflammation in the acute stage of infection. Hot, wet towel packs can also be applied to the affected part in the chronic phase. The incidence of mastitis can be reduced by proper milking followed by covering the ends of the teats with a mild disinfectant.

Mycoplasma is an infection with any one of several mycoplasma organisms. *Mycoplasma mycoides* causes highly contagious pleuropneumonia. Other mycoplasma organisms are referred to as pleuropneumonia-like organisms (PPLO), and these are responsible for a variety of ailments including mastitis, abortion and arthritis including puffy knee joints.

Foot rot can occur when animals are kept on unsanitary, wet surfaces. Lameness of varying severity occurs with a hot swelling of the affected area. This condition can be treated by trimming away all dead tissue and soaking the foot in 10% formalin solution or in a saturated copper sulphate (blue vitriol) solution (0.3 kg/litre of water) for 1 to 2 minutes. Move animals to clean, dry areas. Keeping the hooves trimmed and having animals walk through a formalin or copper sulphate footbath weekly will help control any outbreaks of foot rot.

Brucellosis (Bang's disease) is seen as abortion, lameness, mastitis and/or reduced milk production. Brucellosis is caused by *Brucella melitensis*, a bacterium that causes severe undulant fever in man when it enters the body through a break in the skin. The disease is transmitted through body secretions, including milk, blood, aborted foetuses and placental membranes. Care should be taken to prevent infection when handling any of these items. Goats should be regularly tested for this disease; testing can be made by bacteriological examination of the aborted foetus or of the milk or by the serum agglutination test. This testing is frequently done by government veterinary surgeons. Reactors should be slaughtered to prevent spreading the disease because there is no known treatment. Where brucellosis is endemic, a programme of vaccinating all young animals should be carried out.

Tuberculosis is usually a chronic debilitating disease that affects practically all species of vertebrate animals; the symptoms are similar in various species. Three hardy organisms are responsible: *Mycobacterium bovis, M. avium* and *M. tuberculosis* associated with bovines, poultry and man, respectively. There is often cross infection among species including man. The disease initially shows few symptoms. A general emaciation and weakness with enlargement of lymph nodes develops. Goats (as in man) begin coughing. Post-mortem examination shows lesions (tubercles) developing in various soft tissues of the body. Tuberculosis can be confused with several other diseases such as pleuropneumonia and activo-bacillosis.

Animals become infected by inhaling infected droplets, ingestion of infected discharges, faeces, urine and milk. Close confinement of animals leads to disease transmission.

No reliable treatment exists. Control of the disease depends on the tuberculin test, usually administered by government veterinary surgeons and the slaughter of the reactors.

Anthrax (charbon or splenic fever) is a highly contagious disease to both man and his animals. *Bacillus anthracis*, a spore-forming bacteria, poisons the blood causing fever, oedema, muscle tremors and tissue damage resulting in death from secondary shock and acute kidney failure. In acute cases animals suddenly collapse and have convulsions prior to death. After death, dark tarry blood appears at body openings and rigor mortis does not develop. Animals dead of anthrax should not be necropsied, but burned or deeply (2 m) buried. Disinfection of the area and equipment can be carried out with a 2.5% solution of formaldehyde or chloride of lime.

When exposed to the air, *B. anthracis* develop spores that can remain in animal residues for decades and infect man or animals coming in contact with them.

Anthrax is highly fatal. Early treatment with antibiotics should be given to sick animals. All apparently well animals in the herd and others in the community should be immunised. Annual vaccination should take place where anthrax is endemic. Since anthrax is so highly fatal in man, if its presence is suspected it should be reported to a governmental health agent or veterinary surgeon.

Tetanus (lock jaw) in man and animals is seen as increasing stiffness especially around the head, including the jaws. The third eyelid is protruded. Muscular spasms result from slight stimuli such as touch, noise or light (hyperaesthesia). Animals fall with rigid muscles and die. Temperatures rise rapidly as death approaches. Mild cases may recover after several months.

Clostridium tetani and its spores survive in the soil for many years. They enter the body through surgical and other wounds where necrotic tissue or dirt provide the anaerobic conditions necessary for the proliferation of the contaminating spores and bacteria. They produce a potent poison that passes through the nerves and blood to reach the central nervous system with resulting tetanus. Improper antiseptic surgery can result in tetanus. Since goats are more susceptible to tetanus than most other species except man and horses, it is advisable to vaccinate routinely where tetanus is endemic.

Aftosa (foot and mouth disease) is a serious viral disease among cloven footed animals in many parts of the world. Water blisters (vesicles) develop in the mouth, on the tongue, udder and teats, and between the toes causing soreness, excessive salivation and lameness. Aftosa might be mistaken for vesicular stomatitis.

Goats are not as seriously affected as cattle. Man is generally not affected by the virus. Vaccines have been developed for use in areas where the disease is endemic.

No effective treatment is known. Affected animals should be slaughtered; animals might survive the disease but will never return to productivity.

Caprin arthritis-encephalitis is a contagious retroviral disease that produces severe arthritis throughout the body joints as well as permanent damage to the brain and spinal cord. Some infected animals survive and remain carriers. There is a test to identify carriers, but no known treatment or vaccine. When goats are introduced into an area care should be taken that they are CAE-free.

Abscesses (boils) are quite common in goats. They are often caused by *Corynebacterium* pus organisms as well as *Staphylococci*. Lancing, draining, flushing with an antiseptic solution and systemic antibiotic treatment are effective. The pus that is squeezed out should be buried or incinerated. Caseous lymphadenitis (pseudotuberculosis) causes abscesses to develop in the lymph nodes in various parts of the body. Here they produce body symptoms depending on the organs affected. This disease might be confused with Johne's disease (paratuberculosis).

Contagious ecthyma (sore mouth) is contagious to man suggesting the use of rubber or plastic gloves when treating the lesions and handling vaccines, etc. Blisters (vesicles) should not be opened. A virus causes small blisters to develop on the lips and gums. These rupture and become scabby sores that cover the inflamed, sensitive areas. Infected kids are reluctant to nurse and if they do, they infect the teats of the doe. Invasion of the teat opening causes serious mastitis. Once recovered, the animal is immune. The virus survives over 10 years in scabs. Vaccination should be carried out in areas in which the disease is endemic.

Pneumonia is an inflammation of the lungs by one of the *Pasteurella* or *Hemophilus* species of microorganisms; 'shipping fever' is one aspect. Stress such as high humidity and cold, widely fluctuating climatic conditions, relocation, or an original infection with some other pathogen can all trigger the disease.

The symptoms are high fever, anorexia, depression and respiratory distress including coughing. Death may occur within a few hours to several days after the onset of the disease. Prevention lies in avoiding the stressors and keeping the animals in good shape. Some *Pasteurella* bacterins are effective if administered prior to shipping, introducing new animals to the herd or exposing the animals to other strong stresses.

Nutritional and Physiological Diseases

Nutritional imbalances If an animal is fed enough natural feedstuffs to maintain growth, production and physical condition, it is unlikely that any specific deficiencies will occur. However, in some situations where there are deficiencies of some essential elements in the soil, the plants may grow but will not supply enough of the element to meet the requirements of the animal, and deficiency symptoms occur.

An example of this situation is white muscle disease due to a selenium/vitamin E (tocopherol) deficiency. This problem where the soil is deficient in selenium can be prevented (or treated) by injections of selenium–tocopherol in the pregnant doe or the young kids. Symptoms of white muscle disease include retained placenta, metritis, infertility and heart disease in adults and muscular weakness and stiffness in the kids. Copper, cobalt, manganese and iodine deficiencies are other examples. On the other hand, there are areas where the concentration of certain minerals is so high that animals living in the area become poisoned. Selenium, boron and molybdenum are examples.

Diarrhoea can result from many causes. Digestive upsets frequently occur in milk-fed kids due to irregular feeding (quantity, time, temperature), unsanitary conditions, cold or heat stress, etc., that could reduce the kid's resistance and allow bacterial infection to result. Newborn kids should always receive colostrum which imparts passive immunity to diseases for which the dam has immunity.

Diarrhoea can be treated by removing the predisposing factors and feeding only about half the amount of milk until symptoms disappear. Extra water and electrolytes should be given to prevent dehydration resulting from the diarrhoea. Diarrhoea in older animals can result from excessive

237

succulence in the feed, coccidiosis or other intestinal parasites or other digestive upsets.

Enterotoxaemia (toxic indigestion, overeating disease) produces diarrhoea, depression, lack of coordination, digestive upsets, coma and death. Symptoms result from the toxins produced in the gastro-intestinal tract by the bacteria *Clostridium perfringens* type C or D. These poisons are absorbed into the bloodstream. Predisposing conditions include a high concentrate diet that provides the microorganisms with a type of feed that encourages their growth or anything that could slow down the passage of ingesta through the gut. Enterotoxaemia may occur after excessive feeding by either kids or adults. Animals might overeat after sudden changes in feeds or when under a calcium deficiency or acidosis; hungry goats may overeat if given access to highly palatable feeds that are readily fermentable (like grain). Kids exposed to large increases in milk are at risk. Animals that might be exposed to the predisposing conditions should have been vaccinated with *C. perfringens* toxoid. Enterotoxaemia can be treated with antitoxins.

Enterotoxaemia can best be prevented by offering small frequent feedings whether of milk, grain or forage. Any changes in feed composition should be made gradually over several days. When urea or other non-protein nitrogen is added to the diet, the adaption should be made over a 2 to 3 week period; goats seem to be more sensitive to the harmful effects of urea feeding than cattle.

Acute indigestion with a rumen pH of less than 4.8 indicates lactic acidosis, which can follow high levels of grain feeding in early lactation and could lead to the secondary complication of enterotoxaemia.

The same symptoms occur in acute salmonellosis or in intestinal torsion.

Ketosis (pregnancy toxaemia) results when ketone bodies (such as acetone) accumulate in the body. These compounds are formed when fat is oxidised to provide energy. Normally these fat residues are metabolised (burned) together with carbohydrates. However, if the demand for energy exceeds the energy being supplied by concurrent oxidation of carbohydrates, fats are partially burned and the fragments (ketones) accumulate. Such circumstances occur when animals are off their feed or in starvation and during times of increased energy demands such as late pregnancy or early lactation. Ruminants have a characteristically low blood glucose level; this might be expected in view of the fact that they depend upon the VFA's from the rumen as primary energy sources. The foetus, on the other hand, utilises glucose as its principal source of energy. Since about 80% of the foetal growth occurs in the last 6 weeks of pregnancy, the metabolic glucose requirement during late pregnancy increases by this amount; this only emphasizes the need for proper carbohydrate metabolism. Ketone bodies are acidic and constitute a drain on the alkaline reserve of the body; acidosis develops as a consequence of ketone accumulation. Symptoms of ketosis include loss of appetite, twitching of the ears, muscular spasms, odour of ketones in the breath, milk and urine, frequent urination, going down in a sternal position or lying on one side, coma with rapid breathing and ultimately death. It can be treated in the early stages by drenching with a readily metabolised carbohydrate such as glycerol, propylene glycol or corn sugar and/or with intravenous glucose. Steroid injections are also effective if given after birth (postpartum) but never before birth (prepartum). Ketosis can best be prevented by regularly feeding a balanced, palatable feed and avoiding excessive fatness, especially in females in late pregnancy.

Acidosis can result from carbohydrate engorgement. When grain makes up more than 80% to 85% of the ruminant diet, radical changes in the ruminal microflora and microfauna occur. This changes the pH and the composition of the contents of the rumen which, in turn, can have serious consequences in the animal. Excessive intake of starch or sugar stimulates the growth of bacterial types in the rumen that produce both the D- and L-forms of lactic acid. The L-form is relatively readily metabolised, but the D-lactate tends to accumulate. The pH of the rumen falls from about 6.5 to 4.0 or 4.5 Such acidity destroys the protozoa, the cellulolytic organisms and those that utilise lactate, as well as reducing the rumen motility (Figure 50).

The increase of lactic acid and its salts raises the osmotic pressure in the rumen, which draws liquid from the tissues into the rumen, resulting in dehydration of the body. The acidosis and dehydration can be severe enough to cause death in 1 to 3 days.

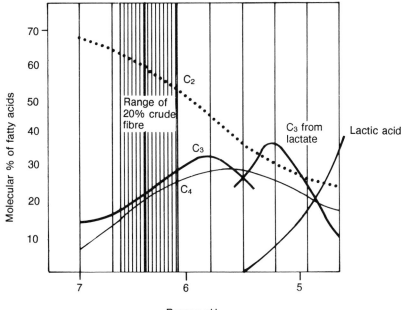

Figure 50   Changes in pH and volatile fatty acids occurring with increased proportions of concentrates in the ruminant ration.

239

Other symptoms of acidosis include reduced milk fat production, depressed appetite, diarrhoea, weakness, erosion of the lining of the rumen, liver abscesses, muscular stiffness, incoordination, laminitis (founder), coma and death. Milk fat depression occurs when more than one-half of the dry matter intake is from grain.

The simplest treatment of acidosis is oral antacids and, where possible, their intravenous injections. However, prevention through feeding adequate roughage is much more successful. Acidiosis in an acute form can result from animals getting into the granary and eating massive amounts of grain. An acute overload of grain can usually be successfully treated by a rumenotomy performed by a veterinary surgeon within the first 18 to 24 hours.

Bloat is seen as a distension of the rumen in the upper left side of an animal. Especially noted is a swelling in the triangle formed by the left hip bone, the end of the rib cage, and the top of the loin. The animal displays apprehension and discomfort. Bloat develops when goats, as well as other ruminants, graze succulent legumes and something causes the eructation (belching) mechanism to fail. The soluble plant proteins, the saponins, and the lack of coarseness in the soft tender leaves, and other factors, team up to produce a viscous fluid in which the fermented gases accumulate, entrapped in a frothy mass within the rumen. The physiology of the ruminant stomach is such that belching (eructation) of gas occurs when sensory nerve endings around the oesophageal opening to the rumen-reticulum are stimulated by the presence of free gas. When these nerves are stimulated by a dryer, coarse material, the regurgitation of a bolus for rumination occurs. However, when these nerves are stimulated by the presence of liquid, the swallowing impulse is stimulated. Thus, when the rumen is unable to separate the gas from the liquid and solid material, the nerves are stimulated by the frothy, wet mass to initiate a swallowing reflex rather than belching; the gas, then, continues to accumulate until it might produce enough internal pressure to restrict breathing and circulation to the point that cyanosis and death follows.

Several forms of treatment for bloat might be applied:

1. Walk the animal about.
2. Stand the animal with its front feet higher than the rear.
3. Tie a round stick in the goat's mouth. This increases saliva production, the mucin in the saliva acting as an anti-foaming agent.
4. Pass a stomach tube (hose) that is 1.3 to 1.9 cm in diameter and 0.6 to 0.9 m in length which might relieve some of the pressure by releasing some of the gas and ingesta.
5. Administer into the rumen some material such as poloxalene, 120 ml of vegetable or mineral oil or household detergent to reduce the surface tension and allow the gas to escape the ingesta.
6. As a last resort, to save the animal, the rumen might be punctured with a trocar and cannula (or knife) physically to release the gas. This puncture must be on the animal's left side and in the centre of the

240

triangle formed by the loin, last rib and transverse muscle (the paralumbar fossa). Adhesions and infections within the abdominal cavity usually follow this procedure. Therefore, it is often advisable to slaughter the animal before these conditions can develop.

Bloat can be reduced or prevented by filling animals with a coarse, dry roughage before allowing them to eat succulent legumes.

Poisonous plants can become a problem when goats are confined to areas where such plants grow. Usually animals avoid these plants but they will eat them if other feed is not available. The list of poisonous plants is lengthy and will vary with the geographical locality. Many ornamental plants and shrubs are included in such a list.

Plants develop natural protective mechanisms, including toxic factors, that retard evaporation and also protect against diseases and being consumed by animals. Goats are less susceptible to some of these factors (like tannic acid) than other ruminants.

Urea toxicity Urea is a naturally produced compound formed in the liver from excess nitrogenous compounds, circulated through the blood to the kidney where most of it is excreted in the urine. Some of it, however, diffuses into the rumen where it becomes a nitrogen source for microbial protein synthesis. Thus, frequently, urea together with a readily available carbohydrate is used partially to replace feed protein. Excessive amounts of urea can result in a buildup of ammonia in the blood to toxic levels. Careful and thorough mixing is required of any feed to which urea has been added. Incomplete mixing can produce pockets of feed that contain excessive amounts of urea. Eating these can be fatal. Urea should supply no more than one-third of total protein in roughage type rations and not more than one-half the total protein in the concentrate portion of the diet. An adaption period of up to 3 weeks is required.

Urolithiasis is confined to males that occasionally develop stones in their urinary tract. Infections in the tract may increase the pH making salts less soluble, and/or provide a focal point for the minerals to deposit and grow. A predisposing factor is an unbalanced ration, particularly one in which there is a high phosphorus content in relation to calcium and/or potassium. The calcium:phosphorus ratio should be near 1.5:1. Ammonium chloride or potassium chloride can be added to the ration to acidify the urine and prevent the crystallisation and deposit of salts.

Milk fever is a result of unbalanced calcium metabolism and is influenced to some extent by heredity. Does with milk fever will have a loss of appetite, excitement, trembling muscles, will fall, go into convulsions, coma and death. Milk fever occurs most frequently in does that have been fed on high calcium legumes prior to kidding when their calcium requirement was low; then, when their calcium requirement increases rapidly as they begin secreting milk, their metabolism does not change quickly enough to keep them from going into an acute calcium deficiency. It is this deficiency that accounts for the symptoms of restlessness, excitement, muscular trembling, falling, coma and death.

Treatment consists of intravenous injections of calcium salts, and since ketosis usually occurs at the same time, glucose is also injected.

Abortion can result when goats suffer from poor nutrition. Diseases such as brucellosis, vibriosis, clamidiosis, viral abortion (that is spread orally) and toxoplasmosis can also be the cause. Abortion seems to be more frequent in older does. Vibriosis and clamidiosis can be prevented by proper vaccination.

Dystocia (difficulty in giving birth) is recognised when delivery has not occurred after 2 hours of straining. Most cases respond to straightening out the one or several foetuses present and rendering gentle traction. Caesarian section is rarely needed but can be successful when performed in time by a qualified veterinary surgeon.

Kid mortality is greatest in the early months of life. Poor nutrition, due especially to near absence of milk, pneumonia and intestinal parasites are the major reasons for low viability of kids. Scours (diarrhoea) is not the major problem with kids as it is with some other young mammals.

SURGICAL PROCEDURES

Castration and dehorning should take place when kids are 3 to 14 days of age. Using a sharp, clean knife, cut off the end third of the scrotum and in turn grasp each testicle and pull it away from the body. If the spermatic cord does not readily separate, it can be scraped in two; a precise cut may lead to haemorrhage. Bloodless castration can be done with the rubber rings of an elastrator or with a Burdizzo clamp, although these procedures are more likely to induce tetanus. In all cases sanitation is necessary to prevent infecting the wound, and where tetanus is a problem, a tetanus antitoxin should be administered at the time of castration.

Disbudding (dehorning) requires destruction or surgical removal of the horn producing cells that surround the horn bud. In the United Kingdom the procedure must be done only by a veterinary surgeon.

The goat horn grows to the skull and has a rich blood supply; consequently dehorning mature animals is a drastic procedure and is seldom recommended. If so, anaesthesia should be administered.

The optimum time for dehorning is at 3 to 14 days of age. The procedure of choice is burning to a copper colour the skin immediately surrounding the horn bud. A stick of caustic soda (NaOH), or several proprietary preparations can also be used to destroy the horn growing cells (Figure 51).

Deodorising can be performed on kids at the time of disbudding by burning or otherwise removing the scent glands that are behind and medial to the horn bud. Clipping the hair on the top of the head of an intact buck will

242

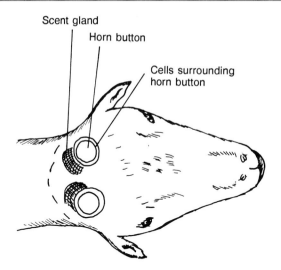

Scent gland

Horn button

Cells surrounding
horn button

Figure 51 *Disbudding and deodorising destroy the horn growing cells surrounding the horn button and the scent (musk) glands of the head.*

reduce the amount of odour at rutting time. Note: does respond to the buck odour with earlier and more intense oestrus. Disbudding, deodorising and castration operations can all be performed at the same time.

Hoof trimming on a regular basis is essential to maintaining goats in a healthy, thrifty condition and preventing lameness and disorders of the feet and legs. Hooves on animals (like fingernails on humans) continue to grow and if not worn off must be cut. Concrete, rocks and gravel cause wear while a wet, soft surface fosters long misshaped hoof growth. The hard hoof can be softened by keeping the animals on wet ground or bedding for several hours before trimming the hooves. The cross-section of the foot shown in Figure 52 depicts the relationship of the hoof to the bony structures. A deep heel is a desirable feature. When looking at the bottom of the foot, the hoof wall should be kept trimmed so it does not either cover the sole or spread outward.

1. Restrain the goat by tying with a halter (or neck chain) to a post or confining in a stanchion.
2. To begin with a front foot, stand at the goat's shoulder and face backward. Press against the body to make the animal shift its weight to the opposite leg. Grasp the lower leg near the pastern, press against the back of the leg to force the elbow to flux as the foot is lifted.
3. With a pointed object (knife blade or farrier knife) clean the hooves. Observe the area between the hooves and under any overgrown and turned over hoof wall.
4. Using the knife, or shears, cut away any folded-over hoof wall. Cut the sidewalls down so they are even with the sole of the foot. If using

243

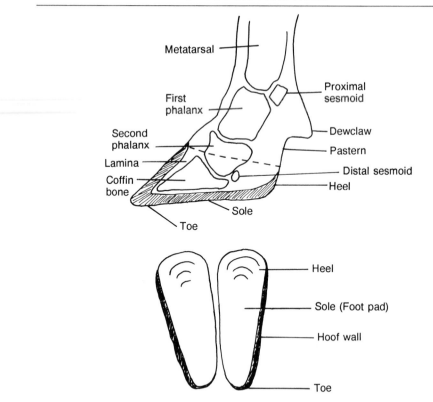

Figure 52   The foot of the goat.

pruning shears, have the cutting blade on the inside cutting outward on the hoof wall.

5. If hooves are too long, cut the tips of the toe back until the white portions within the hoof walls begin to look pinkish. Caution: make several small, thin cuts rather than large ones. If bleeding results from too deep a cut, apply pressure with the fingers to stop the bleeding and apply a tincture of iodine or some other disinfectant. Badly malformed hooves may require several trimmings 2 to 4 weeks apart.

6. To trim the hind foot, with your back to the animal, grasp the dewclaws or pastern and draw one hind leg up between your own, exposing the bottom of the hoof to view. If the goat kicks trying to free its leg, hold it firmly until the kicking ceases and continue the trimming operation.

Caesarean section is rarely needed in goats and should only be attempted by a trained veterinary surgeon.

Devocalising is desirable where bleating creates an undesirable nuisance. The vocal cords can be removed by a veterinary surgeon.

244

12
GOAT LEARNING ACTIVITIES

I SELECTING A DAIRY GOAT

Purpose Learn factors to consider in selecting goats for milk production.

Milk, cheese and other dairy products make palatable and nutritious complements to a vegetable diet. The objective of this activity is to find an animal with the best genetic and physical capacity for milk production. Specialised dairy breeds have been developed which, given the right environment, will produce copious quantities of milk. Under adverse conditions, however, indigenous breeds that have undergone natural selection for survival (rather than production) might produce better.

Production Records
The best indicator of the milk producing capacity of an animal is its past performance. The amount of milk a doe will produce depends on several factors:

1. **Environment** is responsible for about 75% of the variation in production which occurs among animals. This is the reason that management is so important. Animals must be regularly fed adequate balanced diets. They must be protected from diseases, pests, predators and harsh weather. Animals must be handled firmly but kindly and with regularity (since they are creatures of habit). The environment should be such that the genetic capacity of the animal can express itself to the fullest.

2. In developed countries only about 10% of the differences in reproduction among herds is due to **genetics**. Only about 25% of the differences in production among animals within the herd is due to the heredity of the animal.

    The genetic makeup of males is estimated by comparing the production of their daughters with the production of the dams and herd mates of their daughters. Knowing the production of the parents and of the herd, the performance of the offspring can then be estimated. In calculations such as these the average performance of many animals is considered rather than the performance of the individual animal alone.

    As an example: if the parents produced 1000 litres of milk in a herd that averaged only 800 litres annual production, the parents would be 200 litres better than their herd mates. Under the same environmental conditions, the offspring would be expected to produce about one-quarter of that superiority, or 50 litres more than the herd average (or 850 litres). Had the parents produced less than the herd average, the genetic value of the offspring would have been correspondingly less, i.e. 750 litres. The composition of the milk should be considered when comparing the production of different individual does. The richer milk contains more fat and solids-not-fat. For every 1% increase in milk fat there is a corresponding 0.4% increase in solids-not-fat. To make comparisons among animals the 4% fat-corrected-milk (FCM) formula was developed so milk of different fat contents could be equalised on an energy basis.

$$\text{kg FCM} = 0.4 \text{ (kg milk)} + 15 \text{ (kg fat)}$$

Thus, 100 kg of milk containing 3% fat would be equivalent in energy to 85 kg of richer milk testing 4% fat, the calculations being:

$$0.4 \text{ (100)} + 15 \text{ (100} \times 0.03) \text{ kg FCM} = 40 + 45 = 85 \text{ kg FCM}$$

Generally, as the milk becomes richer, total yield declines.

Type or Appearance

The parts of a goat and some desirable aspects of conformation are depicted in Figures 43, 44 and 45, pp. 213, 214 and 215.

Since production records are seldom available, in selection one must rely on some body characteristics (type) that are usually associated with high milk production. They include:

1. A long body with enough capacity to permit the consumption of sufficient feed to sustain life and high production.
2. A large udder with strong attachments holding it up closely to the body. Instead of being firm from fibrous and fatty tissue, the udder should be soft and pliable, consisting of secretory tissue with no hard spots or other signs of mastitis. See Figure 45 for typical udder types.
3. The lactating doe should be free of excess fat, have an angular shape, be sharp over the withers and have flat thighs. The dairy goat should give evidence that any excess feed is converted to milk rather than to fat deposited on the body.
4. The legs should be squarely set on the body, with strong pasterns and a deep heel on the hoof.
5. Freedom from evidences of unsoundness (wounds, injuries, parasites, sickness, etc.).
6. Some genetic factors should be avoided: Hermaphroditism (intersex) is the most important cause of infertility in dairy goats. It is usually related to the polled or hornless characteristic. The genetic female hermaphrodite has a normal sised vulva by 1 month of age. However, the vagina is very short and undeveloped. The abnormality of the genitalia in the male is clearly evident. Congenital hypoplasia in which the genitalia remain in an infantile condition is another genetic factor associated with hornlessness. Cryptorchidism in which one or both testes fail to

descend from the body cavity into the scrotum should be avoided, as also extra teats on either the male or female.

Check your learning Can you:

- Estimate the milk yield of daughters from parents who produced 100 kg more milk than their herd mates who produced 900 kg.
- Determine the 4% fat-corrected-milk equivalent of 2 kg of milk testing 5% fat that a doe produced daily.
- After examining the type of the available animals, identify those most likely to be the best milk producers and those appearing to produce the least.

II REPRODUCING DAIRY GOATS

Purpose Be able to:

1. Mate the doe with a buck.
2. Anticipate parturition
3. Provide assistance in abnormal birth.

Become acquainted with the parts of the reproductive tract of the doe and buck as illustrated in Figures 5 and 8, pp. 17 and 22.

Mating

The goat is a seasonal breeder coming into heat (oestrus) as the days shorten (autumn) to kid after 5 months (145 to 155 days, 150 days average) in the spring. Young does should be bred their first oestrus after reaching 38 to 40 kg in weight (about 10 months) This mating can be delayed if a later kidding would better suit the milk needs of the family. During the breeding season non-pregnant does return to heat every 18 to 24 (average 21) days.

The doe remains in heat for 2 to 3 days; conception is more likely if mating occurs late in the period (bred on the second day). Signs of heat include: uneasiness, riding other animals, standing for riding, shaking the tail, frequent urination, bleating and seeking the companionship of a buck, if present. The vulva may be swollen, red, and some thin mucous may be present. Since an oestrus doe will be a nuisance with other goats, she should be isolated during this time.

For mating it will be easier to bring the doe to the buck rather than visa versa. Only a few seconds are required for the buck to mount the doe and perform the mating act. Remove the doe from the buck's pen after mating has occurred.

Since some bucks may be possessive and aggressive, do not allow them to do damage by butting. By keeping hold of a length of rope attached to the neck of the doe it is not necessary to enter the buck's pen to retrieve the doe.

If the doe has been fed so that she is gaining weight when mated (flushing), the incidence of conception and of multiple births will be increased.

Pregnancy Check

At 100 to 120 days and during the last trimester, pregnancy can often be determined by a technique known as ballottement. Gently force the fist as far as possible into the lower right flank of the doe. This pushes the foetus to the left side out of its normal place. Suddenly release the pressure, then push again. If the doe is pregnant the fist should feel the foetus returning to its position in the lower right side.

Another technique allows pregnancy diagnosis as early as 70 to 110 days in does that are quiet and have been kept off feed overnight and given a soapy enema. A plastic rod about 50 cm (20 in) long and 1.5 cm (0.6 in) in diameter is needed. Or a piece of 1 cm plastic tubing, rounded on one end and stiffened by inserting a metal rod into it, can serve as well. The doe is placed on her back, the fore and rear legs on either side are held together to relax the abdominal muscles. The rod, well lubricated with mineral oil, is inserted about 35 cm (14 in) into the rectum. The front (anterior) end of the rod is moved toward the abdominal wall in front of (cranial to) the pubic bone. Using the other hand, the pregnant uterus can be felt through the abdominal wall as the rod forces it upward. In the non-pregnant doe the end of the rod is felt instead. The empty uterus lies under the udder; it remains small so that it cannot be palpated. The doe should be kept quiet because injuries may occur if she struggles.

Ultrasound and radiographic instruments and increased progesterone levels in the milk or blood serum are more sophisticated methods of pregnancy diagnosis.

Parturition

By 140 days, signs of impending parturition appear. The udder becomes full and hard, with oedema (especially in high producing does in their first kidding). The vulva becomes enlarged and the ligaments around the tailhead relax. At this time the doe should be confined in a clean pen where she will be undisturbed, and where she can be easily caught if she needs assistance.

Symptoms of eminent kidding (parturition) include nervousness, bleating and pawing. After discharge of the cervical plug of thick mucous, the placental sac breaks. Thereafter, the doe should be checked at half-hour intervals, allowing her to give birth without assistance, if possible. Normally the kid is born within an hour. If there is no evidence of progress after she has strained for 2 hours, assistance might be needed.

If assistance is indicated, the doe's movement should be restricted by tying her or putting her in a stanchion. The operator's hands and the genital area of the doe should be cleaned with soapy water and a sanitising agent such as 300 ppm chlorine solution. If the cervix is sufficiently dilated, the lubricated hand can be inserted into the uterus to check the position of the kid(s). The normal position is for the head to be between the front 2 feet. Breech birth can be successful if both hind feet are presented together with the hocks up. If there are 2 legs, be sure that they are either both front or both back. Other positions of the foetus will require its repositioning into one of the two above positions; then give nature a chance to take care of itself. If after another hour birth has not occurred, attach the end of an obstetrical chain or nylon or cotton rope (0.63 cm) to each of the two legs, and then when the doe strains, pull backward and downward, holding the pressure each time. The cardinal rule is to pull only when the doe strains (Figure 46, p. 220).

Upon emergence, first check that the muzzle and nostrils are clear and the kid is able to breathe. Allow time for the blood to leave the placenta and enter the kid through the umbilical cord; sever the cord by scraping or pulling it in two, then dip the kid's navel up to the belly into a tincture (or solution) of iodine in a wide mouth jar. This will discourage disease organisms from entering the body cavity through the umbilicus. If breathing does not commence immediately, hold the kid by the hind legs and swing it in a circle in order that centrifugal force can draw the fluid and mucous from the lungs and air passages. Gentle heart massage and alternating thoracic pressure (artificial respiration) may also be useful. Give the kid access to colostrum as soon as possible.

If the cervix does not dilate or if the afterbirth has not been discharged by 4 hours after parturition, professional assistance from a veterinary surgeon is in order. If such services are not available, the cervix can to some extent be manually dilated, and if the afterbirth has not been shed after 3 days, give assistance in the form of a gentle, slow, steady tension. The cardinal rule in the whole matter of birthing is to be patient, and let nature take its course. Learn beforehand the normal circumstances of time and position so that it can be determined if assistance is indeed needed, and if so, how this assistance can best be given.

Factors responsible for retained placentas include poor nutrition, especially a vitamin A and/or a vitamin E–selenium deficiency, disease such as brucellosis, premature birth and heat stress.

Check your learning Can you:

- Recite the signs of heat in a doe and the season of the year in which they are expected to occur.
- Say how long a doe is expected to remain in heat and when during oestrus she should be mated to obtain the highest conception rate.
- Identify signs of impending parturition.
- Explain the normal and breech presentation of the foetus.
- Recognise the need for assistance at parturition.
- Demonstrate how the doe and operator's hands should be cleansed in preparation of giving assistance at delivery.
- In the absence of an available veterinary surgeon, tell what can be done about a retained placenta.

III MILKING THE GOAT; CARE OF MILK AND EQUIPMENT

Purpose Learn to:

1. Milk a goat.

2. Care for the milk.
3. Wash and sanitise the milk handling equipment.

Milk is not only an excellent source of nutrients for man but for micro-organisms as well. Not all microorganisms are particularly harmful, but there are some that can cause disease and death. These pathogens must be avoided or destroyed before they enter the body.

Milk for food should come only from healthy animals with sound mammary glands. Goat milk is white because the goat is very efficient in converting the yellow carotene in the feed into colourless vitamin A. Strong-flavoured feeds such as onions, turnips and silage should be removed from the goats 3 to 4 hours before milking so that the milk will not be affected by off-flavours.

Milking by hand

1. When properly performed, milking is a pleasant experience for the doe. It should be done regularly twice daily in quiet, peaceful surroundings.
2. Secure the doe either by tying or confining in a milking stanchion. For greatest convenience, a milking stanchion should be 45 cm above the floor, providing both restraint of the doe and a platform on which the milker can sit.
3. Brush off any dirt or particles clinging to the belly and flanks of the doe. With a cloth (or paper towel) wet in warm, sanitising solution, wash the udder and teats with a massaging action.
4. Milk the right teat with the left hand and the left teat with the right. Encircle the teat with the thumb and forefinger, squeezing the teat in such a way as to prevent the backflow of milk from the teat into the gland cistern. Holding this grip, successively squeeze with the third, fourth and fifth fingers applying pressure on the milk within the teat. If enough pressure is applied it will overcome the resistance of the sphincter muscle that closes the end of the teat and will force the milk out and into the pail (Figure 7, p. 20).
5. Draw the first 2 streams of milk from each teat into a container with a clearly visible bottom (strip cup). The presence of flaky, stringy, watery, brownish, thin or bloody secretions is indicative of mastitis. The milk from an infected half of the udder should not be used but discarded in a place distant from the goats. Since mastitis can be transmitted from one goat to another, mastitic goats should be milked last and the hands washed after milking them.
6. To completely empty the udder, massage it downward and while squeezing, remove the residual milk. These strippings will have a higher fat content than the milk previously withdrawn.
7. Dip the ends of the teats in a mild disinfectant.
8. The length of lactation or the time the doe should stop lactating is determined primarily by 2 factors: when she is due to kid and how much milk she is giving. The doe should be dry (not lactating) for about 4 weeks before she freshens. This rest from lactation allows

250

the mammary tissue to become rejuvenated and the body to rebuild its mineral and energy supply.

The milk from some goats becomes salty and rancid near the end of especially long lactation periods. Also the amount of milk secreted may not be enough to be worth the bother of confining and milking her.

9. At least by 4 weeks before the expected kidding date, the doe should stop lactating (be turned dry). Reduce her water and energy intake; skip one milking then stop milking entirely. Glands that have been infected can then be treated with intramammary infusions of antibiotics.

Caring for the Milk

1. After milking, strain the milk through a single service strainer pad in a sanitarily constructed strainer. If these are not available, use a clean cloth previously washed in soap, rinsed first in water, then in a 300 parts per million (ppm) chlorine solution. Straining will remove the coarser foreign matter that might have found its way into the milk.
2. If the milk is to be pasteurised, heat it to 54.4°C for 30 minutes (or in properly designed equipment 73.9°C for 20 seconds). This will destroy any tuberculosis, brucellosis and other disease producing (pathogenic) organisms present as well as many of the fermentative organisms that cause the milk to turn sour.
3. Immediately cool the milk by placing the container in cold water taking care to prevent water from getting into the milk container. Store the milk at 4.4°C and always keep it covered to protect it from flies, dust, animals, etc. The growth of microorganisms is greatly retarded at low temperatures.
4. Milk will heat and cool more uniformly if it is stirred. However, unless care is exercised, stirring can introduce dirt and microorganisms.
5. To protect the flavour of the milk:
   a. Keep the buck widely separate from the does.
   b. If strong flavoured feeds are to be fed (onions, cabbage, silage, young growing cereal plants, etc.) they should be fed only after milking. The does should be separated from such feeds 3 or 4 hours before the next milking.

Cleaning and Sanitising Milking Utensils

The milk handling objective is to maintain the milk (until it is consumed) in the same clean and disease-free condition it was in as it came from a healthy udder. This involves sanitary procedures and prompt cooling.

1. Utensils should have a hard smooth surface. Milk handling utensils should be free from crevices, seams and rust spots that cannot be cleaned. Stainless steel, seamless utensils are preferred, but a well-tinned (free from rust) surface is satisfactory. In improperly cleansed utensils milk will form a base for the buildup of mineral plaques called

'milk stone'. These plaques provide a home and nutrients for the growth of microorganisms.

2. Immediately after use, rinse buckets, strainers and other milk utensils with warm water.
3. Wash utensils in hot water to which has been added an alkaline detergent according to the manufacturer's directions. With a brush remove all traces of soil from the surfaces. Once weekly replace the alkaline detergent with an acid cleaner in the washing water to remove and prevent mineral deposits from forming on the surfaces.
4. Rinse utensils in hot water and then drain them in a place free from dust, flies and other insects. Exposure to sunshine has a sanitising effect. Microorganisms cannot grow on a clean, dry surface.
5. Before use, sanitise utensils by rinsing them in a chlorine solution of 300 pm. Household bleach can be a source of chlorine. Other chemical sanitisers can be equally effective when used according to manufacturer's instructions. All sanitisers are effective only on clean surfaces, free from organic matter.

Check your learning   Can you:

- Milk a goat after using a strip cup.
- Prepare milk for storage.
- Clean and sanitise utensils for handling milk.

IV   FEEDING GOATS

Purpose   Develop growing, lactating and maintenance rations for goats.

When born, a kid is essentially a simple-stomached animal depending (as any other mammal) on milk to supply the amino acids, vitamins and other essential nutrients. Later, many essential nutrients are provided by microbial activity in the mature rumen. Gradually, as more and more roughages are consumed, the rumen develops so that after weaning the animal is capable of living on a solely roughage diet.

The goat is a herbivore with a definite requirement for fibre. It cannot live indefinitely on grain alone. However, fibre in a feed can fill the available gut space, thus restricting the nutrient intake. For maximum milk production the high yielding doe cannot consume sufficient roughage to meet her nutrient needs. She requires some concentrates to meet those needs for production. Concentrates in addition to roughage are needed to obtain optimal growth of weanlings, also.

The moisture in feedstuffs can also fill available space in the gastro-intestinal tract; the moisture content of pasturage, green chop and silage can restrict the amount of dry matter (and nutrients) that can be consumed.

Roughages
Roughages are classified according to the moisture content and the harvesting and storage methods as pasture, green chop (soilage), silage and

hay. Young plants with the leaves attached will have the highest protein content and digestibility. Pasture and green chop will have essentially the same composition, 80% to 90% moisture, except that the animal in harvesting its own feed (pasturing) will be somewhat selective. Silage contains 60% to 70% moisture and is packed into a silo to exclude air to retard mould growth. The fermentable carbohydrates in the stored forage are converted to lactic and other acids which retard the growth of microorganisms that would otherwise reduce its feeding value. Hay represents the entire plant dried to less than 18% moisture; leaf loss reduces the feeding value of hay. The nutritive value of roughages varies greatly with the plant species, stage of maturity, harvesting and storing methods and conditions.

Although classified as a browser, goats are effective grazers as well, depending on the type of forage available. Goats are rather selective in their feeding behaviour, selecting the leaves, flowers and fruit that contain more protein and less fibre. Thus they select feeds that are more digestible than the stems and petioles. Goats will select grasses when the protein content and digestibility of the grasses are high. Then they switch to browsing when the overall nutritive value of browse may be higher. Their small mouths and prehensile lips and their agility and ability to stand on their hind legs give them a distinct advantage in foraging. When fed a poor quality feed such as straw, the goat is at a disadvantage because there is no opportunity to be selective. It has a smaller digestive tract in relation to its body size than does a cow and so does not utilise poor quality roughages as well.

Grazing increases the energy requirement of a goat due to the time spent looking for and getting to the feed. The extra energy expenditure can increase the maintenance requirement by 15% to 60%. A grazing goat on the average spends about 30% of its time feeding. Depending on the particle length of the forage, another 30% is spent ruminating. The 2 primary feeding periods are from daylight to mid-morning and from 3 hours before sunset until darkness.

Daily total dry matter intake of lactating goats varies from 3% to more than 5% of their body weight.

All animals should have ready access to clean water.

Access of the goat to a half and half mixture of dicalcium phosphate and trace mineralised salt is also desirable; loose salt is generally preferred to a salt block. The specific minerals needed for supplementation will depend on their relative lack in the local soils and water. Vitamin A (carotene) can be obtained from green feed.

Concentrates

Concentrates are more readily digested than roughages. They include the grains maize, barley, wheat, sorghum grain, soybeans, cottonseed, quinoa and products such as cassava chips, coconut cake, soybean meal, cotton-seed meal and fish meal. When fed, the grains should be cracked, crushed or rolled to disrupt the impervious hull. This exposes the inside of the seed to the action of digestive enzymes. Dustiness in feeds should be avoided.

Feeding the Animal

Animals require nutrients for maintenance and for production (growth, gestation or lactation). The maintenance requirement is a function primarily of body size. The larger the animal, the greater will be its maintenance requirement. Requirements for extra activity, maintaining body temperature, growth and gestation are usually expressed as a factor of maintenance to be included in the maintenance requirement. Of all body operations, lactation has by far the greatest demand for nutrients. The requirement for lactation will vary with the amount of milk produced and the concentration of fat and other solids in the milk. Water and a salt lick should be available to the animal at all times.

Feeding yearlings and the breeding buck   Browse, good pastures and/or high quality hay made up of legumes or part legumes and a salt lick will provide the nutrients needed for growth. A calcium–phosphorus (e.g. dicalcium phosphate) and other necessary mineral supplements are recommended especially if the roughage is grass.

If animals become too thin they should receive 0.4 to 0.7 kg grain mixture daily. Avoid excess flesh; if the ribs cannot be readily felt, the animal is too fat.

Feeding the pregnant dry doe   Less persistent does will cease lactating by themselves but all does should have 4 to 6 weeks of rest from lactation. To dry off a persistent doe, reduce her intake of water, remove the grain from her diet and feed only poor quality roughage and after 1 day on this feed, skip 1 milking, then stop milking her entirely. After 2 days, return the doe to full water intake and maintain her on a diet of good quality roughage. The milk in the udder will be resorbed.

If the roughage is only lucerne it should be changed to grass for the last 2 weeks of gestation. This precaution reduces the risk of parturient paresis (milk fever) because the non-lactating doe, with a minimum calcium (Ca) requirement, is challenged to rid herself of the excess Ca when on a diet of Ca-rich legume hay. Then, after kidding, when she begins secreting Ca-rich milk, her Ca requirement is drastically increased. She cannot alter her metabolic processes fast enough to prevent a Ca deficiency (milk fever).

Feeding the lactating doe   About 230 kg of hay and 200 kg of grain are required to sustain a lactating doe for a year. Mediocre production can be supported by free choice good quality roughage alone, but increased production requires some concentrate. High producing goats in temperate zones are fed a ratio of roughage to grain of:

40/60 in early lactation
50/50 in mid lactation
60/40 in late lactation.

Goats should never be allowed to become fat; this can be a matter of concern in late lactation. Forage dry matter can be fed daily at 2% to 3%

body weight. The concentrate can then be relied on to provide the balance of the needs. The concentrate should contain 12% to 14% protein if the roughage is legume or 16% to 18% if grasses provide the roughage.

In tropical areas, a simple supplement for milk production fed to does on pasture might be by weight:

44% sorghum grain
37% maize grain
10% soybean meal
6% molasses
1% urea (mixed in well; too much urea can be fatal)
2% salt and mineral mix.

On good pasture, silage or hay, milking does usually require only 1 kg of grain for each 4 kg of milk. On only fair roughage, supplementation should be 1 kg of grain per each 2 kg of milk.

The efficiency with which animals convert feed energy into animal products varies with the physiological activity. Maintenance is more efficient than growth; fattening is the least efficient. Lactation is comparable to maintenance; in fact, if a doe is lactating, she lays on fat with the same efficiency as milk secretion. The energy loss in heat increment varies with the type of feed, roughages producing more heat increment than do concentrates. Digestible energy (DE) and total digestible nutrients (TDN) do not take this into account. Consequently these measures of feed energy overevaluate the energy values of roughages as compared with grain. For this reason, net energy becomes the measure of choice when grain is substituted for roughage (and vice versa) as energy sources.

Rules of thumb and approximations will often provide all the information desired. However, where more precision is desired in calculating the nutritive needs of animals, reference should be made to one of the many available tables of requirements and feed composition. The estimated net energy (ENE) value for a feedstuff can be estimated from TDN values by the formula:

Mcal ENE/100 lb = 1.393% TDN – 34.63

Note that as the energy concentration increases the relative value of the negative constant (34.63) becomes less in the equation.

With the information in Table 2 and Table 12, rations for goats of different sizes in different physiological conditions can be calculated. As an example, with early cut lucerne hay, maize and soybean meal available, what should be fed to a doe weighing 50 kg and producing 2 kg of milk containing 4% milk fat? To reach a solution, prepare a table such as shown at top of page 256.

The amount of a feedstuff needed to supply a specific need can be determined by dividing the amount needed by that in 1 kg of feed. Thus, 1.21/1.76 = 0.68 kg of maize will meet the energy deficit of 1.21 Mcal ENE when maize has 1.76 Mcal/kg. At 9.3% protein, 0.68 kg of maize would provide 63 g of protein, which exceeds the requirement of 14 g; consequently, soybean meal would not be needed. The maize would also supply 0.68 kg × 0.28% = 1.94 g of phosphorus, but since this is less than the

255

	ENE Mcal	Protein g	Calcium g	Phosphorus g	Carotene mg
Requirements for:					
Maintenance (50 kg)	1.34	91	4	2.8	4.5
Production 2 kg @ 4%	1.40	144	6	4.2	19.0
Total	2.74	235	10	7.0	23.5
Feed					
Lucerne, 1.25 kg (@ 2.5% of body weight)	1.53	221	16.6	2.9	52.0
Concentrate must supply the difference	1.21	14	surplus	4.1	surplus

4.5 g needed, a phosphorus supplement such as dicalcium phosphate should be supplied.

Another way the values in Tables 2 and 12, pages 44 and 224, can be used is in answering such questions as how much barley grain supplement should be fed daily to meet the extra needs of late pregnancy? Using another measurement system, Table 12 shows an additional need of 397 g TDN. Barley is 74% TDN or 1 kg of barley will supply 740 g of TDN. Therefore, 397/740 = 536 g of barley will be needed to supply the energy required for late pregnancy.

In solving problems such as these, it is well to remember that excesses of other nutrients that come in meeting the energy needs are usually of no concern. In the milking doe example, had grass hay been fed, an additional protein (as in soybean meal) would have been needed. When the tabular values do not fit the circumstances, interpolated values or those next higher ones can be used. Also, if values are not given for a particular feed, use those for a similar one (e.g. use sorghum values for millet).

Retain only animals for which there is sufficient feed   When feed supplies are short there is often a reluctance to reduce the size of the herd even though by doing so the total production would increase. Animals can produce only after they have energy above that needed just to stay alive. As an example, if a farmer produced only 14 kg of early growth lucerne hay each day, he could maintain 10 goats, but these goats would not have enough feed to produce any milk. Any milk they might actually produce would be at the expense of their body reserves. However, if 3 of these animals were slaughtered or sold, the maintenance overhead for the remaining 7 would be reduced to 9.8 kg leaving 4.2 kg of hay available to produce 6.1 kg of milk. If a total of 5 animals were sold, 7 kg of hay could produce 10.1 kg of milk. Thus, when feed supplies are limited, more total production occurs with fewer animals.

Feed Analysis

For the past century knowledge of the composition and digestibility of feedstuffs has depended largely on the proximate analysis that was developed at the German Weende experimental station. The feed fractions, the procedures followed and the principal components of these fractions are outlined on page 50.

The availability of fibre for energy depends on microbial fermentation in as much as the animal does not have any enzymes of its own to break down the hemicellulose and cellulose. Differences exist in the digestibility of the crude fibre fraction from different feedstuffs. Grasses have a high content of cellulose in their cell walls and low lignification which leads to a high digestibility by ruminants. On the other hand, legumes such as lucerne have low cell wall contents but these cell walls contain a high percentage of lignin. Since lignin is not fermentable by rumen microbes this leads to a low digestibility of the crude fibre fraction in legumes.

A more recent and precise analytical process for measuring the fibre of a feedstuff is the Van Soest method (Figure 17, p. 51). Here different detergents followed by sulphuric acid and burning are used to separate the easily digested cell contents from the fibrous cell walls (neutral detergent fibre). The fibre is then separated into hemicellulose, cellulose and lignin components. The neutral detergent component is probably the best predictor of digestibility of a feedstuff by ruminants.

Check your learning Can you:

- Identify the factors that influence the nutrient needs for maintenance.
- Determine the amount of early bloom lucerne hay needed to provide the energy required for maintaining a kid weighing 10 kg.
- Determine the amount of sorghum grain needed to offset the additional energy requirement of growing 50 g per day.
- Using locally available feeds, develop an adequate ration for a lactating doe weighing 30 kg and producing daily 1 kg of milk containing 3% fat.

V SANITATION AND HYGIENE FOR GOATS

Purpose To understand the need for sanitation and disease prevention among animals.

Diseases cause illness and death. Healthy animals are free from disease.

Look through the microscope and see some of the tiny forms of life that are present but cannot be seen with the unaided eye. Most of these do not hurt us; in fact, many are beneficial. Some of these organisms, however, make us sick (or cause diseases). Those forms of life that are harmful are called pathogens. They are of several different forms such as viruses, bacteria, moulds and protozoa. Diseases might even be caused by worms and insects.

There are many diseases among animals. Only a few examples are pneumonia, aftosa, tuberculosis, brucellosis, anthrax and enterotoxaemia. Some of these diseases like tuberculosis (in cattle but not in chickens), brucellosis and anthrax can be transmitted to man. Climate, soil, lifestyle and other factors influence the types of diseases that might be common to an area (endemic).

Pathogenic organisms are so small that they cannot be seen without the aid of a microscope; they are called microorganisms. In order to cause damage, the pathogens must first enter the body through the digestive tract, respiratory system, intact or injured skin, or other body openings. Pathogens are transferred from sick to healthy animals by such means as manure, body secretions, coughing, physical contact, and flies and other insects. Flies breed in warm, moist organic matter. For this reason, manure and discarded feedstuffs should not be allowed to accumulate but should be gathered and spread on the fields or gardens for fertiliser.

To prevent sick animals from making the healthy ones sick too, they should be kept separated. When new animals are brought onto the farm, or if animals are returned after being away, they should be kept separate for 3 to 4 weeks (quarantine) so that if they get sick they will not contaminate all others.

Wild birds and animals frequently carry disease organisms which, if they come in contact, can cause the disease in domestic animals. Therefore, wild birds and animals should be kept away not only from the farm animals but also from their feed and pens.

If the domestic animals are well fed, protected and healthy, they are more resistant to the pathogens. An animal's natural resistance to some diseases can be increased to higher levels by vaccination.

In addition to invisible pathogens, internal and external parasites can also cause damage to animals, and therefore must be controlled. These parasites include protozoa, roundworms, tapeworms, flukes, mites, ticks and fleas. The purpose of sanitation is to destroy or weaken the pathogens and parasites before they can harm the animal. Chemicals that destroy pathogens are called disinfectants or sanitising agents. Anthelmintics and insecticides are used to control worms and insects.

Inadequate sanitation is one of the most important causes of health problems in goat production. Sanitation begins with cleanliness. Manure, rotting feed, hair and rubbish should not be allowed to collect, but should be either tilled into the land for fertiliser or burned. It is wise to move the kids' pens periodically to clean ground.

Pens and cages in which sick animals have been kept should be thoroughly cleaned and disinfected before other animals are placed in them. Also feed and watering devices should be cleaned and disinfected after each period of use. Scrape away caked dirt and manure and wash with a brush and water before rinsing with a disinfectant.

Disinfectants can be prepared by several methods and should be applied to clean surfaces. Manure and other dirt can neutralise the effects of disinfectants, because the disinfectant reacts with the dirt rather than the pathogen; therefore, the equipment should be cleaned before the disinfectant is applied. Disinfectants are more effective when applied hot. Some effective disinfectants are:

30 g or ml sodium hypochlorite (laundry bleach) dissolved in 1 litre water.

1 g lye (NaOH) in 45 ml water.

1 ml cresol per 32 ml water (a detergent will facilitate emulsification).

When animals are kept confined on the same land for long periods of time or when large numbers of animals are concentrated on small areas, disease organisms and internal and external parasites build up in numbers and virulence. The pattern can be broken by moving animals to fresh uncontaminated ground or buildings and by tilling the ground.

Check your learning Can you:

- Name the types of organisms that cause disease.
- Distinguish between pathogens and internal and external parasites as causes of disease.
- State how a disease is communicated from one animal to another.
- Explain how quarantine can help control disease.
- Explain why wild birds and animals should be kept away from domestic animals and their feed.
- Explain why equipment and surfaces must be cleaned before applying a disinfectant.
- Identify at least one disinfectant.
- Describe how to break the buildup cycle of pathogens and internal and external parasites.

VI GOAT HEALTH AND DISEASES

Purpose To become acquainted with the diseases that goats are most likely to encounter and learn how they can be recognised and controlled.

Diseases are conditions that impair normal functioning. They may be due to a variety of causes (aetiology) which will influence the symptoms, severity and any type of treatment necessary for their cure.
  A few animals in isolated situations with freedom to cover wide areas are not subject to many diseases. This situation changes when many animals are maintained in small areas and contact is made among several herds. Infections are also more prevalent in warm, moist areas.
  Under intensive production situations goats have been found to be susceptible to the same diseases that affect cattle and sheep.

Internal parasites (helminths) such as roundworms, stomach worms, whipworms, lungworms and tapeworms propagate through eggs laid by mature worms in the stomach or intestine of the goat. The eggs are voided with the faeces and then hatch in the earth into infective larvae which crawl up stalks of forage to be consumed by a new host goat. The young worms mature in the digestive tract, lay eggs and the process is repeated. The cycle can be broken by pasture rotation and/or by consistently treating the goats with a deworming compound (anthelmintic) such as thiabendazole or ivermectin.

Mastitis is an inflammation of the udder caused by one or more of many factors. Infections can follow udder injuries, improper milking, infection

259

with a virile organism or a drop in general resistance. The udder often becomes sore, swollen and hard and the milk flaky, watery or tinged with blood. Milk with an abnormal appearance can be detected with a strip cup or by passing a thin layer of milk over a dark coloured surface. A loss of secretory tissue with a consequent drop in milk production is an almost inevitable consequence of mastitis. Treatment involves making the animal comfortable, milking to keep the udder empty, infusion of antibiotics into the udder or systemic treatment by intramuscular or intravenous injection of antibiotics. Or, if antibiotics are not available, cold water and ice packs may reduce swelling and inflammation in the first stages of the disease. Hot towel packs followed by milking and gentle massage 4 to 5 times daily might be practised in the chronic stage.

Brucellosis symptoms in goats are abortion, lameness, inflammation of the udder and reduced milk flow. This disease in goats is caused by *B. melitensis* which readily causes undulant fever (Malta fever) in man. It is thought to be transmitted to man through unpasteurised milk, but a more likely infection route is through sources of pathogens such as aborted material, body excretions or meat of infected animals coming in contact with a break in the skin.

Bloating is the accumulation of excessive amounts of gas in the rumen. It can result from quickly eating succulent legumes which contain saponins and high levels of soluble proteins. These factors produce a stable froth that encapsulates the gas as it is formed in fermentation within the rumen. Treatment consists of breaking up this froth by such things as increasing the flow of saliva which contains mucin, or drenching with a household detergent, mineral oil, etc. As a last resort to save the animal's life, a trocar and cannula (or knife) can be used to puncture the rumen on the left side in the centre of the triangle formed by the left hip bone, the end of the rib cage and the top of the loin. Unfortunately, this procedure results in a good deal of the frothy ingesta leaking in between the walls of the rumen and abdominal cavity causing infections and adhesions. It might be best to butcher such animals as soon as convenient.

The incidence of bloat is greatly reduced if animals are filled with dry coarse feed before being put on leguminous pastures.

Ketosis is a metabolic disease that occurs following some stress such as going off feed, being transported to a new location or kidding. It is more common in fat than in thin animals and in does carrying multiple foetuses. Symptoms include a twitching of the ears, muscular spasms and loss of appetite. The breath and urine (and milk) will have the odour of acetone. If unchecked, coma can develop with rapid, laboured breathing, frequent urination and finally death.

Treatment consists of drenching with 0.5 to 1 litre of propylene glycol or molasses or an intravenous injection of glucose. Feed the animals a balanced, palatable diet.

Milk fever (parturient hypocalcaemia) is seen as a loss of appetite followed by restlessness, excitement, trembling of the muscles, falling, convulsions, coma and death. The disease results from a deficiency of calcium in the blood resulting from the loss of calcium from the body when heavy milk secretion begins and when the endocrine system has not had time to adapt to the increased need. Predisposing conditions of milk fever are a high calcium, low phosphorus prepartum diet such as in legume hay as the only feed. A high phosphorus, low calcium diet for the 1 to 2 weeks prior to kidding can lessen the incidence.

Milk fever is treated by intravenous injections of calcium salts, usually a calcium gluconate which also treats the concurrent ketosis.

Check your learning Can you:

● Name and describe symptoms and control practices for:
  a. Internal parasites.
  b. External parasites.
  c. Infectious diseases.
  d. Nutritional and physiological diseases.

Special emphasis should be placed on those problems that are commonly encountered in the area.

VII MANAGEMENT OF GOATS

Purpose To become familiar with production management techniques.

In handling goats it should be observed that, as contrasted with sheep, they are more individualistic with no strong herding instincts. They are creatures of habit, curious and affectionate. Their successful management includes several handling techniques.

Raising Kids

Kidding Kidding has been discussed in detail in Learning Activity II, Reproducing Dairy Goats. Kidding should take place in a dry, clean area that in cold weather is bedded and protected from draughts. The doe should be where she could easily be caught if need be. Normally, immediately after birth the doe will lick the kid. This not only cleans the hair coat but also stimulates the kid, giving it increased interest in getting up and enthusiastically beginning life's course. If for some reason the dam does not do this, the operator should clear the nostrils and dry off the kid with clean rags or towelling.

As soon as possible, the kid should receive colostrum. If the kid does not suck, the colostrum should be milked out and fed through a nipple. Cleanliness should be observed especially when feeding milk. Colostrum is a concentrated source of protein, minerals and vitamins, but more important, it imparts passive immunity to diseases in the newborn kid.

Examine the kid for extra teats, wry face, hermaphroditism or other obvious abnormalities. The wisdom of spending time and effort on deformed kids is questionable.

Males grow faster than females.

Feeding milk The first month of life is very critical to kids. Milk feeding is a transition providing nutrients to the animal until a functional rumen develops. The frequency, amount and type of milk feeding will influence the growth, health and rumen development in growing kids. The objective is to develop the rumen and wean the kids from milk as soon as possible, leaving as much milk as possible for family and market use.

When the kid is born, the abomasum (true stomach) makes up about 80% of the total stomach space. This proportion is reduced to 8% to 10% in the adult stomach (Figure 12, p. 27). There is an oesophageal groove beginning at the end of the oesophagus and traversing to the abomasum while communicating with the rumen, reticulum and omasum. The oesophageal groove, when closed, by-passes the forestomach and transfers fluids directly from the oesophagus into the abomasum. This closure is an autonomic action prompted by a happy feeling that is characterised by eagerness to eat and a rapid wagging of the tail. Nursing from the teat or a nipple bottle is more likely to induce this reaction than drinking from a pan. Things such as quickly gulping the milk, milk of a different temperature, or if the kid is uncomfortable or ill-at-ease, prevent the reflexive closure of the oesophageal groove, the milk may then overflow into the rumen where it only sits and ferments because there are no enzymes or other facilities in the rumen with which to digest and absorb it.

Frequency of feeding:

1. Feeding milk ad libitum results in most rapid body weight gain, increased milk consumption and retarded rumen development. Under such conditions kids drink up to 25% of their body weight daily before 4 weeks of age and 15% thereafter.
2. Feeding only once daily is not satisfactory because it fails to stimulate enough stomach and intestinal secretions to properly digest the milk and prevent intestinal disorders from an overgrowth of microorganisms in the upper intestine.
3. Feeding kids 3 or 4 times daily will result in higher intake and faster gains but after 6 days of age feeding twice daily is generally adequate.

Quantity: growth performance of kids is linearly correlated with the intake of milk; the more milk consumed the faster the gain. As noted above, kids given free access to milk can consume up to 25% of their body weight daily; however, feeding more than 10% of the body weight uses milk that could otherwise be available to the family. When milk is overfed the digestibility of other nutrients is reduced. Too much milk can cause scours and stomach upsets. Dry feed intake (grain and roughage) is encouraged by restricted milk feeding. The intake of dry feed stimulates the development of bacteria and protozoa in the forestomach; the organic acids, especially acetic, that result from microbial growth are necessary for the proper development of the rumen including the papillae. There is a tendency to overfeed young kids and underfeed older ones.

Age of kid (days)	Milk Feeding Rates for Hand-Fed Kids Number of feedings/day	Maximum kg daily per kid
1–3	4	Colostrum, ad lib.
4–5	3	0.45
6–10	2	0.68
11–Weaning	2	0.90–1.3

Milk replacer can successfully replace goat milk after the kid is 2 weeks of age if it is economically feasible to do so. The dry matter content in the replacer should be 16% to 24% fat and 20% to 28% protein. Replacing animal fat and protein with that from plants reduces kid performance. A crude fibre content of over 1% is indicative of plant components. The dry matter in the reconstituted milk can vary from 12% to 24% (fresh milk is 12% to 14%). The replacer as fed day to day should have a constant concentration because varying the proportions of powder and water leads to digestive upsets. Any extra colostrum, if diluted up to one-half so that it will have the same solids content as milk, can replace a part of the milk. Sugar (sucrose) should not be fed to kids because they have no enzyme (sucrase) with which to digest it. Sucrose in the feed leads to digestive upsets.

Milk or milk replacer has been successfully fed cold; in fact, if milk is fed free choice it should be cold (4.4°C) to discourage overconsumption. However, feeding cold milk requires a higher level of management skill. Generally, best kid-raising results come from feeding milk at a uniform 38.9°C to 40.5°C.

Regularity in feeding is very important. This applies to amount, temperature and composition of the milk as well as time of feeding, environmental conditions, etc.

Always wash and sanitise nipples, bottles, pans or any other utensils used in feeding the kid. Keep troughs, water buckets and mangers clean.

Feeding water and dry feed When 1 week old, kids should be supplied with fresh, clean water and fed a dry grain starter, then after a few days a good quality roughage should be added.

Weaning Kids can be successfully weaned at 9 kg body weight (or 2 to 2.5 times birth weight), 8 weeks of age or at the time they are consuming 30 g/day of solid feed.

1. Weaning is a critical phase of kid management. The kids should weigh about 2.5 times their birth weight at weaning. Efforts should be made to

increase the intake of solid feed so that they have reached that weight in 8 weeks.

Early weaning, shock and coccidia infestation pose hazards at weaning. The 30 to 50 g of dry matter to be consumed should consist of both concentrate and good quality roughage.

2. Other factors affecting weaning include variables such as sex, breed of goats and the physical forms of diet. Females are more resistant to weaning stress than males and Alpines more than Nubians. Before weaning, kids should be fed both hay and grain. Kids grow faster if they are fed grain only, but development of the reticulo-rumen depends on the intake of roughage. Weaning shock is reduced if the general health status of the herd is at a high level.

Weaning to breeding   The object of raising the dairy goat kid is to develop a producing animal with an adequate body size in the shortest time (1 year) and with the least cost. Due to the lack of development of the digestive system, the immature goat must receive some concentrate to supply the nutrients not provided in the roughage. The better the quality of roughage, the less concentrate is required. A good quality, leafy lucerne hay fed in sufficient amounts to allow the animal to do some selecting plus a daily supplement of about 0.9 kg of grain should adequately provide for optimal growth. Access to fresh water, salt and trace minerals is necessary. Internal parasites (especially coccidiosis) need to be controlled. Where this is a problem, treatment with a coccidiostat should begin at 3 weeks of age and continue at periodic intervals until bred.

Animals reach puberty when they have achieved about one-third their mature weight.

All goats are active and need facilities on which to exercise. Blocks or other objects on which they can climb and jump are suitable.

Hoof Trimming

Frequently, the goats do not have opportunity to wear down their hooves, which then require frequent inspection and trimming as described earlier on page 243.

Disbudding

The horn grows at the poll from a ring of special cells surrounding the horn. Disbudding is the process of destroying these cells in the young kid before the horn develops. If discoloured skin is fixed to the skull in two rosettes, horn buds are present. Moveable skin, however, indicates a naturally polled (hornless) condition.

Before a week (3 to 5 days) of age, and if the goat is healthy, heat an iron hot enough to 'brand' a piece of wood, hold the kid firmly by the muzzle and press the iron to the horn button for 10 to 20 seconds or until a bronze colour has been produced over the horn producing cells. Repeat with the other horn. No further treatment should be necessary. The success of this process cannot be assured if the horn has already grown over 1.25 cm in length (Figure 51, p. 243).

Dehorning removes the horn and the ring of cells from which it develops. Due to the way a goat's horns grow onto the skull, dehorning after horns have grown beyond 2.5 cm (1 inch) is a difficult surgical procedure.

Buck kids can be partially deodorised by destroying the scent glands that are located along the inside and back of the horn buttons. This can be accomplished concurrent with disbudding.

Examine kids for extra teats and remove these with a pair of scissors. Such animals should not be used in a breeding programme because of the undesirability of more than 2 teats. Supernumerary teats is a condition that is hereditary.

Castrating

Only males from outstanding parents should be kept for breeding purposes; they should remain intact. Buck kids to be slaughtered under 2 months of age need not be castrated. However, all other males should be castrated before they reach 2 months (preferably 1 to 3 weeks) of age. One method of castration involves the use of a sharp knife. Wash, sanitise and dry the hands, knife and kid's scrotum. Have a helper hold the kid by the hind legs, the goat's back to the helper's chest. Grasp the bottom of the scrotum with one hand, stretch it tightly and cut off the bottom third of the scrotum. This will expose the ends of the testicles. Grasp 1 testicle at a time, pulling it from the body and severing the spermatic cord. If the kid is over 1 month of age, it might be necessary to scrape (not cut) the cord in 2 with a knife. Protect the surgical site from flies and infection and the wound should heal with no further attention.

To prevent infection, all kids should be vaccinated with 0.5 to 1 ml tetanus toxoid at the time of all surgical and other wounds and punctures. Vaccination should be repeated within 3 to 4 weeks.

Identification

The identity and parentage of newborn kids should be recorded. This not only makes it possible to distinguish the kid but will establish ownership as well. Tattooing in the ear or notching the ears is generally preferred over branding.

Administering Medicine Orally

1. **Drenching** To administer up to 1 litre of liquid:

   a. Put the dose of medicine in a long-necked bottle.
   b. Back the goat into a corner and encircle its neck with the left arm.
   c. Insert the left thumb into the goat's mouth behind the incisors but in front of the premolars and force its mouth open far enough to insert the neck of the bottle 3.8 to 5 cm into its mouth.
   d. Tilt the goat's head slightly upward and slowly pour the liquid over the rear of the tongue.

   e. Do not pour so rapidly that the goat gasps for air and breathes some of the medicine into the lungs.

2. In the absence of a balling gun, to administer a bolus or dry medicine that has been securely wrapped in paper, hold the animal as above but pull the tongue outward to one side. On the opposite side insert the bolus with the fingers and push it over the back of the tongue, then as soon as the fingers have been removed release the tongue. The goat will swallow the bolus.

3. Exert caution that the sharp teeth and strong jaws of the goat do not do harm to the fingers.

Check your learning   Can you:

- Trim the hooves on a goat.
- Disbud a kid.
- Castrate a buck kid.
- Drench a goat.
- Explain the importance of receiving colostrum within the first 2 hours of a kid's life.
- Explain why regularity in time, amount and temperature is important in feeding milk to a kid.
- Give the conditions of age, body weight and dry matter consumption necessary to reduce weaning shock in kids.

VIII   GOAT DAIRY PRODUCTS

Purpose   Learn to process milk.

Milk is a palatable source of high quality protein, vitamins and minerals, as well as energy. It can be consumed as it comes from the doe or it can be processed into a variety of products that can be stored for later use. The milk protein will coagulate to form a curd when exposed to an increase in acidity and/or to certain enzymes such as rennin.

   The flavour that develops in a fermenting dairy product is influenced by:

   the types of microorganisms present;
   the substrate or food on which the microorganisms live, which might be the milk sugar (lactose), the fat, or the protein;
   the end products formed;
   the temperature and length of time over which fermentation takes place;
   and the amount of other flavouring agents added (salt, sugar, seasonings, etc.).

Milk

Milk can be consumed as the fresh liquid. Its care and pasteurisation were discussed in Learning Activity III. The milk should first be pasteurised for the most consistent results in making other products.

Buttermilk

Milk can be fermented to buttermilk by adding a starter in the form of 10 ml buttermilk per litre, or by adding a commercial mixture of micro-organisms. These ferment the lactose into lactic acid and ferment certain other milk components to impart desirable flavours.

Fermentation should take place overnight at 21°C to 24°C. Some prefer the taste imparted by the addition of a dash of salt.

Cheese

Cheese can be made either from whole milk or from partially skimmed milk. Many types of cheese are made depending on:

the source of milk;
the content of fat, protein, salt, moisture and any added ingredients;
the fermenting microorganisms;
processing procedures;
length of the aging period;
storage conditions;
temperatures used in processing, etc.

It takes about 10 parts of milk to produce 1 part of cheese. During the cheesemaking process the milk divides into the liquid whey and the solid matter curd. The whey contains most of the water, milk sugar (lactose) and water-soluble vitamins, and some of the protein, fat and minerals. It can be used in cooking (if the family is lactose-tolerant) or fed to the livestock or processed further to yield a speciality cheese. The curd will contain about three-quarters of the original protein, much of the fat and fat-soluble vitamin A, and about half of the minerals.

The following cheesemaking procedure is one that can be altered to fit the family preferences and conditions of manufacture and storage:

1. Beginning with 2 to 4 litres of pasteurised milk in a large kettle, adjust the temperature to between 30°C and 31°C. Add 1% to 2% buttermilk, or other starter, to produce flavour and acid and allow to set at 31°C overnight. Adding rennin reduces this holding time to 30 minutes. (Rennin is the enzyme produced in the stomach lining of a young mammal such as a calf or kid. Its purpose is to coagulate the milk as it reaches the stomach to hold it so that digestion can begin.)
2. The curd is firm enough to cut when it will break clean over the index finger inserted at an angle and gently raised. Using a long knife, cut the curd into approximately 1.3 cm cubes by making 3 cuts at 1.3 cm intervals, 1 diagonally, 1 at right angles to that cut, then followed by a vertical cut crosswise to the previous cuts. (See illustration.) Slowly reheat (0.3°C /minute) to between 37.8°C and 40.6°C. Gently stir occasionally to keep the curd from lumping together. Hold under these conditions until the curd is firm. At this point the curd will have shrunk to about one-half original size, and the curd will squeak between the teeth as it is chewed. This process will require 2 to 3 hours. The higher

267

the temperature and the longer the heating, the drier (firmer) the resulting curd will be. If the curd is not firm enough, the cheese will be pasty
and may sour. If the curd is too firm, the cheese will be dry and crumbly.

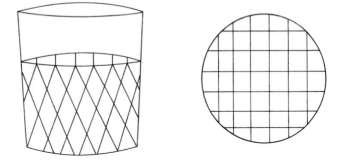

3. When the desired consistency is reached, pour the curd and whey
   into a large container (colander) lined with cheesecloth. Lift out the
   cheesecloth and the curd that has been caught in it. Let it drain a few
   minutes, rinse out additional whey with cold water, add salt to taste
   and press overnight.
4. A simple cheese press can be made by punching outward holes in the
   bottom and sides of a can large enough to hold the curd. Cut a round
   1.25 to 2.5 cm thick piece of wood slightly smaller in diameter than the
   can. Place a clean cloth in the bottom of the can. Add the curd that has
   been caught in the cheesecloth along with the cloth itself. Cover the
   curd with the excess cheesecloth. Place the round board on top and
   apply pressure equivalent to the weight of 6 to 8 bricks to the board.
   With appropriate equipment the pressure is increased gradually over a
   few hours up to 23 kg, which pressure is held overnight. In the absence
   of a can, the salted curd, still in the cheesecloth, can be shaped into a
   ball. Wrap its circumference with a folded cloth pinned in place with a
   safety pin. Place the ball between 2 boards and apply pressure (as with
   6 to 8 bricks) on the top for 1 or 2 days. This cheese must be turned
   several times a day. Generally, the longer the cheese is pressed and the
   greater the pressure applied, the harder will be the cheese.
5. When the pressing is completed, remove the cheese from the press,
   remove the cheesecloth, wipe the cheese with a clean dry cloth and
   place it in a cool, (13.3°C), dry place. The cheese should be turned
   daily. Drying at the surface will produce a rind. After this rind is
   formed, paraffin the cheese by melting 225 g wax, dip half the cheese
   in the wax, remove and allow to cool. Then dip the other half in the
   same way. Cover any cracks or bubbles with additional wax; mould
   will form on any curd that is not covered by wax. Return the cheese to
   13.3°C storage, turning it daily. Mild cheese can be eaten after 6 weeks,
   but usually cheese is aged for 2 to 6 months for a stronger flavour.

Butter

Due to the small size of the fat globule in goat milk it is more difficult to get the fat to rise so it can be skimmed off. However, if a mechanical cream separator is not available, put the cooled milk in a shallow pan and allow to set undisturbed; skim off the cream after 12 to 24 hours. Adjust the temperature of the cream to between 8.9°C and 12.8°C, place in a churn or tightly closed jar and agitate until the fat coalesces into globs. Churning is the phase-changing process whereby the continuous water phase (with interspersed fat globules) found in cream is changed to a continuous fat phase (with interspersed droplets of water) found in butter. Drain off the liquid buttermilk, wash the butter with cold water working with a wet, wooden paddle, mix in enough salt to taste, and place in a covered container for cold storage until consumed.

If frozen and sealed, butter can be kept for many months without any decline in quality. If unprotected, butter will absorb off-flavours to which it is exposed. A less desirable storage method is submerging 0.2 to 0.5 kg chunks of butter in a saturated salt (NaCl) solution.

Check your learning Can you:

• Prepare a fermented milk drink.
• Make cheese from goat milk.
• Make butter.

IX GOAT MEAT FOR FOOD

Purpose Learn to slaughter goats for meat.

Excess males and older non-productive animals can be slaughtered for meat (chevron or cabrito). The following relate to this food and how it might be processed:

1. Goat meat is characterised by a lack of fat both under the skin (subcutaneous) and interspersed in the muscles (marbling). Excess fat in goats tends to be deposited more among the viscera.
2. A carcass that weighs about one-half the live weight can be expected.
3. Kids can be butchered at birth and dressed like rabbits.
4. Castrated buck kids (there are generally 115 males born for every 100 females) fed for 6 to 8 months provide perhaps the choicest meat. Disbudding and deodorising at 10 days improves the flavour of the meat.
5. Older animals will be tougher and have a stronger goat flavour. This flavour is minimised in mature bucks by castrating them and removing scent glands 2 months before slaughter. Meat from older animals can be made into jerky, salami or other products where tenderness and mild flavour are not especially critical.
6. Empty the gastro-intestinal tract by fasting the animal for 24 hours before slaughter.

7. Dispatch the animal with a gun or stun it with a sharp blow from a heavy hammer on the skull above and between the eyes, over the brain. Cut the jugular veins in the throat and allow thorough bleeding by hanging the carcass, head downward. The blood can be collected and if not consumed by the family can be fed to chickens or pigs.

8. Suspend the animal by its hind leg(s). To skin, slit the skin the length between the hind legs, over the midline of the abdomen to the throat and out along the inside of the forelegs (Figure 53A). Once the cut is started, the fingers can be inserted under the skin holding it away from the body for faster cutting. Cut around the anus, pull out a length of the rectum and tie off with a string; cut off the skin at the base of the tail. Beginning at the hocks, carefully separate the hide from the carcass without cutting it, cutting as close to the skin as possible. The skin should be stretched out flat to dry for subsequent tanning. Sever the head at the base of the skull.

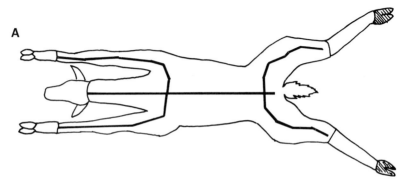

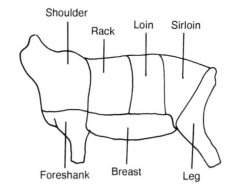

Figure 53   Skinning and cutting a goat carcass.

270

9. Open the body cavity by cutting down the midline taking care not to puncture the gut. Tie off the bladder, if possible. Allow the paunch and intestines to roll out. Loosen the colon end down past the kidneys and carefully remove the bladder. Remove the liver and cut out the gall bladder including a piece of the liver. Tie off the oesophagus ahead of the rumen; cut the oesophagus allowing the offal to fall free. Split the brisket using a meat or wood saw if necessary, and remove the diaphragm, heart and lungs. Remove any remaining pieces of tissue, wash the carcass with cold water and wipe dry with a clean cloth.

10. The tongue can be cut out, and the skull can be split to remove the brains. The liver, heart and tongue should be washed in cold water and set to dry.

11. The surface of the freshly slaughtered carcass is contaminated with bacteria that can spoil the meat unless their growth is checked. Chilling is the method of choice. The goat carcass is not protected by a thick layer of subcutaneous fat as is a steer, for example, so does not age as well as some other species. Wrapping the carcass with a clean cloth will partially protect it from contamination.

12. The wholesale cuts of a goat are represented in Figure 53B. Begin cutting the carcass by removing the thin cuts—the breast, flank and foreleg. Remove the kidneys and kidney fat and the neck. Separate the carcass into 4 prime cuts. A cut between the fifth and sixth ribs removes the shoulder. Another cut between the twelfth and thirteenth (last) rib separates the rib. The loin and legs are separated just in front of the hip bones by cutting through the small of the back where the curve of the leg muscles blends into the loin. Split the legs apart by cutting through the centre of the backbone.

Further cutting can be carried out to meet the family needs. Mature goats will usually yield from 22% to 28% boneless meat based on unshrunk live weight.

13. Due to its low fat content, relatively low temperatures and moist cooking techniques should be used when cooking goat meat.

Check your learning   Can you:

● Butcher a goat for meat.

13
CAVY (GUINEA PIG, CUIS) PRODUCTION PRACTICES

Cavies, commonly called guinea pigs, originated in South America. They are small, tailless rodents with 4 digits on the front feet and 3 on the hind ones. Cavies are said to be members of the order *Rodentia*, family *Caviidae* and genus *Cavia*. There are at least 5 species in this genus. The domestic cavy is known as *Cavia porcellus*.

The *C. cutleri* was domesticated by Andean Indians long before Pizarro conquered Peru around A.D. 1530. The guinea pig forms a part of the diet of natives of the Bolivian, Ecuadorian and Peruvian high country, where the animal, after being killed, is scalded to facilitate the removal of the fur by scraping; the eviscerated carcass is then wholly either fried, roasted or added to a stew. Cavies are also reportedly a part of the human diet in Camaroon, Africa.

SELECTION

For the breeding herd guinea pigs should be selected that are large for their age with broad shoulders, have brilliant eyes, sound teeth, sleek haircoats and are active and full of spirit. Females (sows) live for about 5 years and males (boars) 6 to 7 years. The body parts of a guinea pig are shown in Figure 54.

REPRODUCTION

Males are somewhat slower to mature than sows, reaching puberty at 8 to 10 weeks of age and ready for service at 3 to 4 months (weighing 600 to 700 g).

Females reach puberty at 4 to 5 weeks of age and can first be bred after 2 months (weighing 350 to 450 g) but should be bred before 6 months. In non-pregnant, non-lactating sows, oestrus occurs every 15 (range 13 to 20) days and lasts for up to 50 hours. They ovulate 10 hours after beginning oestrus. Should there be a sterile mating the female usually exhibits pseudopregnancy.

Gestation lasts 65 days (59 to 72 day range) with a litter size range of 1 to 6, averaging 2 to 3. Sows have only 2 teats that are located in the inguinal

272

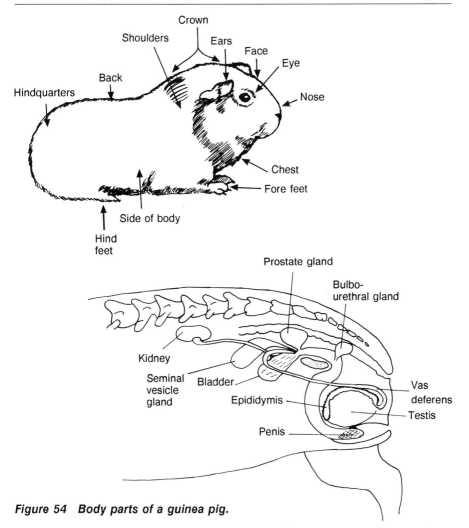

Figure 54 Body parts of a guinea pig.

region, yet are known to have raised 3 to 5 offspring. Sows might produce 3 to 5 litters per year. They double their weight during pregnancy.

Abortions are independent of litter size, but the stillbirth rate is proportional to the litter size. Stillbirths occur more often in the first litter a sow bears. About 8% of pregnancies terminate in abortion.

Birth and rearing The sow will generally mate within 2 to 3 hours postpartum with an 80% conception rate if bred at that time. Otherwise, open sows return to heat only after weaning the litter.

The weight of individuals at birth depends on the nutrition of the

273

sow, genetics and litter size; within a litter of 3 to 4, the young weigh an average of 85 to 95 g each. Young weighing less than 50 g at birth usually die.

Newborn cavies are relatively large and precocious: they are furred, and have open eyes and erupted teeth. The young begin to move about immediately after birth, and they begin eating solid feed when 1 to 2 days old. The sow sleeps apart from the young. Guinea pigs have a gregarious nature and do well when grown in colonies.

In Latin America the young are weaned at 10 to 14 days when they weigh 200 g. Others may choose to wean the young at 3 to 4 weeks when they weigh from 165 to 240 g. Their development is rapid, gaining 2.5 to 3.5 g/day during the next 2 months, after which time they should weigh 370 to 475 g. In colony situations, to prevent the sire from mating with his daughters, one or the other must be removed before the young are 33 days old. The males and females should be separated before 2 months of age. The females live in peace, but the males fight among themselves.

Guinea pigs reach maturity at 5 months but continue to gain weight for up to 12 to 15 months, at which time breeds used in experimental laboratories might weigh 700 to 850 g for the sows and 950 to 1200 g for the boars. In South America maturity is reached with 1000 to 1300 g for females; males weigh 200 g more. Efficiency of reproduction declines after 27 to 30 months of age.

It is relatively easy to distinguish the sex. While holding the animal, belly away from the operator and with the genitals between the thumb and forefinger, press down gently on the abdomen. This forces the penis to emerge from the sheath of the males.

Mating Systems

Monogamous pairs (one female per male) require maintaining many more boars, but the system produces a larger number of offspring per pregnancy.

Polygamous groups (several sows per boar) require less space per animal maintained and have the further advantage of a higher incidence of postpartum matings and of communal rearing of the young. Each sow requires 1160 sq cm (180 sq in) floor space, which space is more important than the number of sows per boar (up to 4 or 5 per boar). Because of the aggressive nature of the males there can be only 1 boar per pen.

FEEDING

1. **Feedstuffs** Cavies are vegetarians eating hay, weeds, seeds, fruits and vegetables. Guinea pigs do not efficiently convert feed into body tissue. They possess a small stomach. Fibre fermentation occurs in the caecum and large intestine; efficiency of fibre digestion is much like a horse. Guinea pigs do practice coprophagy. In contrast to other animals, vitamin C must be included in the diet, but in common with other herbivores, there must be some fibre also. As with any herbivore that must depend on microorganisms for fibre digestion, rapid changes

in diet should be avoided. An intake of only highly succulent feeds can produce diarrhoea; however, some greens are needed to supply the dietary vitamins C and E and also to enhance the ability of the animals to maintain their body weight. If no green feed is fed, it will be necessary to supply 10 mg vitamin C per kg of feed or 200 mg ascorbic acid per litre of drinking water. Colonies receiving greens in addition to their pelleted feed weaned 10% to 44% more offspring than those not receiving greens.

2. **Protein**   Guinea pigs have a high requirement for some amino acids which is met by a vegetable diet containing as little as 14% protein. Raw soybeans can be fed to guinea pigs. The trypsin inhibitor in raw soybeans does not seem to affect the digestive process.

3. Cavies should be fed twice daily, but if fed only once, this should be at night when the animal naturally does most of its eating. In any case, only the feed the animals will clean up should be provided. Mouldy feed should be avoided. For maximum growth, some grain is required.

   The animals make a noise when hungry.

4. **Water**   Guinea pigs are often raised in their native land without being provided with any free water. The water they receive is from succulent feeds and that derived from metabolism. As a general rule, however, as with other animals, clean water should be available. One recommendation is 5 to 100 ml per day for animals receiving a green supplement, and 250 to 1000 ml for those on dry feed only. Watering devices should be so designed and secured that the water remains available and is not spilled.

MANAGEMENT

Cavies have poor eyesight but excellent hearing and smell. They are creatures of habit and respond to regularity. Their development of rigid habit patterns makes changes in feed, water, housing, etc. stressful. Guinea pigs are gentle and respond to frequent handling, thus making good pets.

In handling a guinea pig, pick it up by the shoulders or top part of its body and slide the other hand under the body with its chest and forelegs resting on the wrist. Damage to lungs and liver can result from grasping around the thorax or abdomen.

Being nocturnal in nature, cavies need at least 10 hours of darkness for best results. Natural mating occurs at night. Direct sunlight should be avoided.

The cavy tolerates cold better than heat. The ideal temperature is 15.5°C to 23.9°C (60°F to 75°F) and relative humidity 50%. Heat prostration and death can result from exposure to 32.3°C (90°F) for extended periods. The area in which the animals are maintained should be dry and well ventilated, providing 10 to 15 draughtless air changes per hour.

Clipping the toenails of cavies kept in cages is sometimes necessary.

275

Pens

Cages for a 250 to 300 g cavy should provide a minimum of 374 sq cm (58 sq in) floor space and 17.8 cm (7 in) height; 2 sq ft of floor space would be preferred. The cage floor should have openings 1 cm apart to allow droppings to fall through although a solid floor with clean bedding has been found more satisfactory for long-term rearing. A 318 mm hardware cloth covered by a coat of shavings permits the urine to run through, keeping the pen dry. Since guinea pigs do not jump or climb, if the cage sides are about 30 cm high, a top is generally not required to retain the animals.

A perch and hidden runway add to the comfort and feeling of security of the animal. When excited, pigs in colonies tend to stampede, trampling the young. Using rectangular pens or cages with barriers can reduce circling.

A nest box 30 cm long, 15 to 18 cm wide and 15 cm high can be provided in which the pregnant sow can give birth to her young. Birthing can occur successfully on solid pen floors.

HEALTH

Guinea pigs are highly susceptible to the toxic effects of many of the common antibiotics, particularly when given orally. The problem seems to come from the inhibiting effect that the antibiotics have on the normal intestinal flora on which herbivores are so dependent. Antibiotics remove competition and enhance the overgrowth of *Clostridium perfringens* and *C. difficile*, resulting in enterotoxaemias. In particular, those antibiotics with activity against Gramm positive bacteria (penicillin, lincomycin, and erythromycin) should not be administered orally. Furthermore, antibiotic ointments should not be used in treating skin diseases because they are ingested by the animals as they lick themselves. There are some broad-spectrum antibiotics such as gentomycin and kanamycin that, if injected, can be used with caution in treating pneumonias and other bacterial infections.

Cavies should have clean, well-ventilated but draught-free quarters. They are messy housekeepers requiring that their cages, pens and feeders be cleaned at least weekly. The use of detergents and disinfectants with an occasional weak acidic (vinegar) solution to reduce urine scale and crystals is recommended.

Colds and pneumonia Guinea pigs are highly susceptible to respiratory diseases including the human cold and *Bordetella* pneumonia transmissible from dogs, cats, rabbits and rodents. For this and other reasons, cavies should not be housed with other species. Symptoms of respiratory diseases include eye and nasal discharges, excessive lacrimation, overly brilliant eyes, laboured breathing, hovering in corners and body temperature of 40°C to 41.1°C (normal 37.7°C to 39.4°C). With few treatment options available, the usual treatment is to make the animal comfortable and provide palatable feed.

Diarrhoea is seen in the animal as voiding liquid faeces, having a rough haircoat and refusing to eat. Diarrhoea can result from various digestive upsets such as excessive amounts of highly succulent feed, or from the presence of internal parasites. Salmonellosis causes a severe form of diarrhoea and high death loss.

Other infectious diseases include internal and external abscesses and infections caused by *Staphylococci*, *E. coli* and *Corynebacteria* organisms. Fungi (ringworm), which is also contagious to man and other animals, can be a problem.

External parasites afflicting the cavies are:

1. **Lice**—small wingless insects of both the biting and sucking varieties.
2. **Fleas**—small wingless blood-sucking parasites.
3. **Mites**—by burrowing into the skin these minute, rounded or oval, short-legged, flat organisms produce mange or scab which causes severe distress and agitation. Animals infested with these external parasites show skin irritation, itching, loss of hair and condition. Lice and fleas can be seen with the naked eye but a magnifying glass is needed to see the mites.

Control of these parasites lies in cleaning and disinfecting the pens and dipping the animals in an insecticide such as 0.5% malathion, paratox or cuprex.

Internal parasites Intestinal coccidiosis (*Eimeria caviae*) and hepatic coccidiosis (Tyzzer's disease) are serious internal parasitic (protozoan) infections that can be controlled by proper sanitation.

Scurvy The guinea pig is the only farm animal requiring dietary vitamin C. Scorbutic animals show unsteady gait, painful locomotion, haemorrhage from the gums, have swollen, painful joints, rough haircoats and loss of weight.

Muscular dystrophy Guinea pigs are very sensitive to a vitamin E deficiency. They become stiff and lame and refuse to move. Vitamin E is found in green leaves and oils from seeds. A selenium deficiency can cause similar symptoms.

Pregnancy toxaemia (ketosis) may occur in obese sows or those going off feed prior to parturition. Affected cavies do not eat or drink, but have muscle spasms and coma within 48 hours of onset. This condition may cause death within 4 or 5 days unless parturition occurs to change the physiological status.

Congenital dental problems include malocclusions and overgrowth of incisors and molars.

Common causes of death include embryonic resorption, uterine haemorrhage, pregnancy toxaemias, dystocias, and exhaustion from prolonged labour.

Good husbandry is essential to good health and this includes the practice of quarantining new animals for at least 2 weeks before putting them with the home animals.

14
PIG PRODUCTION PRACTICES

Pigs are omnivorous and in some respects compete with man for food, yet these animals can use many by-products and wastes derived from human feeding. Since they have litters and gain rapidly and efficiently, they can provide a quick return for the money invested. Pigs have not only provided man with meat and fat but pigskins, bristles and manure as well. The manure is valuable as fertiliser for the soil or for fish ponds. For these reasons pigs are found almost everywhere except where religious laws restrict the eating of pork. Pigs do not always require man's attention; in many areas they can live by themselves (feral) if given the opportunity. Feral pigs can be trapped or hunted for food.

SELECTION AND BREEDING

Calorie conscious consumers in the developed world have forced a change in hogs away from the shorter, more compact meat-lard type (as in the Hampshire) to the longer, thinner bacon type as seen in the Landrace and Yorkshire breeds. Also, a carcass with less fat is obtained by slaughtering at a smaller body size and weight.

Animals for breeding should be free from genetic defects and be from families noted for large litters and early sexual maturity. They should be healthy, have sound feet and be well grown for their age. Since baby pigs until weaning depend largely on the sow as a source of nutrients, the sow needs to have the milk-producing capacity to nourish her litter; this will require at least 12 uniformly spaced normal teats. For retention in the herd, the sow should demonstrate her ability to produce large, fast growing litters.

Boars should display the same type of characteristics as the females (including 12 evenly spaced teats) except that he needs to display male (rather than female) qualities. Since only one boar is needed for many females, a much greater selective pressure can be applied to the breeding males; in fact, selection for body type, rate of gain and feed efficiency can be optimised in male selection.

REPRODUCTION

Female pigs come into heat at 21 day (19 to 24 day) intervals throughout the year and stay in heat for up to 48 hours. Oestrus is characterised by grunting, restlessness, seeking to be with the boar, swelling of the vulva and becoming immobilised with ears erect when pressure is placed on her rump. Bigger litters are obtained if the sow is bred during mid or late oestrus, the time at which ovulation takes place. This, together with the large number of ova shed by the sow, suggests that if the sow is to be bred only once, this event should occur in the morning of the second day of heat. Better results, however, will follow if the sow is bred on the first day of heat, then re-bred 12 to 24 hours later. Some herdsmen recommend using a different boar for the second service. Copulation lasts from 5 to 20 minutes.

Gilts should be first bred on or after their third heat when they are 6 to 8 months old and weigh 102 to 113 kg. A sow will not cycle when lactating but will come into heat 3 to 7 days after her pigs are weaned. She can be re-bred at that time if in good condition.

The gestation length is 114 days (3 months, 3 weeks and 3 days). With proper management this makes it possible to produce 2 or more litters per year.

Boars can breed at 7 to 8 months of age but do not become completely sexually mature until 15 months. If in a thrifty condition, he can ejaculate up to twice daily for 3 to 5 days. With proper management, a boar can serve as many as 50 sows; however, additional boars will be needed if the sows are grouped such that several are to be bred within a few days.

Since a boar can serve up to 50 females each year if matings are properly spaced timewise, it is costly and wasteful of resources for each producer of 1 or 2 sows to maintain a boar. However, with cooperative use of a boar among several neighbours, there is always the danger of transmitting disease from one farm to another. Artificial insemination could answer this problem if there were sufficient technological capability. Unfortunately, at this time, sperm in boar semen cannot survive freezing. Semen is copious (300 to 500 ml), and for best results, a large quantity (100 ml) of diluted semen containing 2 billion sperm needs to be inseminated into the sow. Properly collected and diluted semen can be stored at 16°C to 18°C for up to 2 days. A village artificial breeding cooperative among those raising pigs might be considered.

The strong flavour of body tissue makes boar meat unpalatable. To make this meat more acceptable, the boar should be castrated 30 to 60 days before slaughter.

Infertility may result from genetic abnormalities, poor nutrition, disease and adverse climatic conditions. Pigs do not have sweat glands. During hot weather neither boars nor females are as active, leading to missed heat periods. In the male high temperatures can result in temporary loss of fertility. In the female extreme heat causes early embryonic death and reduction in the number of pigs born alive and surviving. Also, the gestation period is shortened by high temperatures. On the other hand, newborn pigs are very susceptible to cold, being unable to regulate their body temperature for the first few hours of life; pigs are a few weeks old before developing this capacity.

FEEDING

In the past and in many tropical villages today, pigs tend to be free to roam at will as they scavenge, cleaning up human and animal faeces and other offal as they can. Although feed costs might be minimal with this type of management it is also associated with malnutrition and infestation with internal parasites. In more intensive pork growing operations, feed cost can account for 80% of the total production cost of pork, justifying attention to this aspect. If properly fed and managed, the pig can grow very rapidly from 1.2 kg birth weight to 100 kg 6 to 8 months later.

Feedstuffs

Forage Pigs can obtain only a part of their feed needs from forage. Balanced diets containing 20% to 40% sun-dried lucerne meal can be fed successfully to growing-finishing pigs. The structure of the mouth, teeth and snout permits feeding on the surface of the soil or rooting feed out of the ground. In the humid tropics it is difficult to manage pigs on pasture or on the ground due to the internal parasite problem. Green forage is beneficial to penned pigs; adults can eat up to 4 to 5 kg per day. Such feeding can reduce the need for vitamin and mineral supplements.

Grain Of feedstuffs available for feeding pigs, the cereal grains usually provide the basic energy source; corn is generally the standard of reference. However, the usual varieties of corn are deficient in lysine and tryptophan followed by threonine and the sulphur-bearing amino acids, and total protein. Other cereal grains can be fed depending on availability and relative costs. Sorghum and millet have about 95% of the feeding value of corn and can make up 35% to 50% of the total ration. Their protein is, like corn, deficient in some essential amino acids. Wheat is similar to corn in energy value but has a higher protein value.

Barley and oats have a higher protein content, but a lower energy feeding value than corn due to their higher fibre content. All the seeds should be crushed or coarsely ground before they are fed.

About 15% of the diet should come from a protein supplement. Besides protein, grain rations are likely to be deficient in calcium and some vitamins and will need to be supplemented with these factors.

Roots and tubers can be successfully fed at up to 30% to 40% of the total ration. It is advisable to cook cassava before feeding in order to destroy the toxic cyanogenetic glycosides found in the skin of some varieties. Potatoes, beans and soybeans also should be cooked for pigs.

Table scraps and other human waste and even faeces can be fed to hogs with varying results. Four pounds of heavy waste is considered equivalent to one pound of grain. Transmission of disease is one deterrent to feeding waste. Municipal waste is cooked at 65.5°C for 30 minutes to help control trichinella (causes trichinosis in man), vesicular exanthema and other diseases of pigs and man. Cooking reduces the ability of the pig to sort the feedstuff; since the pig would eat only the most digestible portions, if given the opportunity, cooking reduces the digestibility of the total ration, especially crude protein.

Pigs incorporate dietary fatty acids into their own body fat. Since waste is high in unsaturated fat, if fed to hogs it will produce a soft, oily carcass. This is one disadvantage of feeding it to hogs.

Feeding Management

1. In designing a ration the energy requirement must first be met. The protein source must supply not only general nitrogenous needs but also special organic structures, the essential amino acids. Proteins from various sources (vs. single source) are more likely to supply all of the essential amino acids; especially is this so if the proteins are from animal sources and/or soybean meal or other pulses. Maize and soybean meal make a happy combination to supply energy and protein.

   The nutritive needs of pigs vary according to age, weight, sex, stage of gestation or lactation and environment. However, the most critical periods are (a) birth to 18 kg weight, (b) last trimester of gestation, and (c) lactation.

2. If vitamin–mineral premixes are not available, all classes of pigs should receive some fresh, succulent forage daily. For minerals, some fresh, clean soil or grass sod can be put in the pens although this practice carries the danger of introducing internal parasites.

3. Generally, in developed countries a complete mixed feed is before the animals at all times, allowing them to eat when and how much they choose. Exceptions are non-lactating sows and boars that might become fat unless the feed intake is reduced or is diluted with some low energy materials such as high fibre roughage.

   Where facilities for mixing complete feeds are lacking, pigs should be hand fed the various quantities of different ingredients; pigs cannot balance their own ration even if all necessary ingredients were available to them.

4. Rapid increase in body weight up to 45 kg will encourage muscular tissue growth, but after achieving this weight, further rapid live weight increase encourages the formation of fat.

5. Pigs should have drinking water available at all times. In the tropics a lactating sow may drink up to 28 to 35 litres daily.

DISEASES AND PARASITES

There are many recognised diseases of pigs. The pig producer should be generally aware of their causes, symptoms, treatment and control. It is not expected that the owner of a few pigs should know of every abnormality, yet it is important to be able to recognise when animals are sick and when to seek professional help from any available veterinary surgeon or animal pathology laboratory in the area.

Contagious Virus Diseases

Enteric viruses that cause diarrhoea:

1. **Rotavirus enteritis** generally infects the small intestine of nursing pigs that are 1 to 6 weeks old. It is also known as white or milk scours. The infection and diarrhoea caused by rotaviruses resembles that seen in coccidiosis and enzootic transmissible gastroenteritis (TGE). It is not as serious as TGE. Dehydration is a dangerous side effect. There is no effective vaccine or treatment available; prevention depends on sanitation, good husbandry including colostrum and milk at an early age, keeping pigs comfortable and isolating the sick.
2. **Transmissible gastroenteritis (TGE)** causes vomiting and diarrhoea in pigs of all ages, but mortality is higher in the young. No highly effective vaccines are yet available although a number are of some value and help reduce mortality. Strict sanitation, isolation, and good husbandry are necessary in controlling this virus.
3. **Adenovirus, enterovirus** and **pararotavirus** are other contagious viruses that cause diarrhoea.

Vesicular diseases

1. **Foot and mouth disease** (aftosa) is an acute, extremely contagious viral infection (*enterovirus* and *picornavirus*) of cloven-footed animals, still present in many parts of the world including Africa, Asia, parts of Europe and South America. It is very debilitating with lameness, and vesicles (blisters) of feet, tongue, mouth, nose and internal organs. There is no effective treatment. Vaccines are effective in prevention.
2. **Swine vesicular disease** (SVD) (*picornavirus*) is very similar to foot and mouth disease and is transmitted through uncooked food waste and raw pork products fed to pigs. Its diagnosis depends on serological tests.
3. **Vesicular exanthema of swine** (VES) (*calicivirus*) is transmissible from sea lions to pigs and African green monkeys. Symptoms are the same as foot and mouth disease. **Note:** when any of the vesicular diseases develop, your local governmental health agent should be notified. Quarantine, test and slaughter, and strict sanitation (disinfect with alkali or hypochloride) should be practised.

Other viral diseases

1. **Hog cholera** (swine fever) is an acute, highly contagious, highly fatal pig disease endemic in many countries. The first signs of an acute outbreak include dullness, reduced appetite, constipation changing to diarrhoea and a fever peaking at 41°C to 42°C about 4 to 6 days after onset. The skin over the snout, ears and abdomen often becomes cyanotic. Haemorrhagic lesions develop on the inside lining of blood vessels and other organs. Cholera has been eradicated in the USA and in a few other countries via vaccination. Hyperimmune serum is used in treatment. Antibiotics are of no value in treating this or other viral diseases.

283

2. **African swine fever** (ASF), caused by *iridovirus*, is highly contagious and usually fatal. It is present in many parts of the world including Africa, Brazil, Portugal, Spain and Italy. It is not found in the USA, most of Europe, Cuba, Haiti and the Dominican Republic. Symptoms resemble those of cholera. The virus is spread in all fluids, tissues and excretions of all acutely infected pigs. No effective vaccine is yet available. Eradication and control come through stringent quarantine, test and slaughter programmes. All suspicious cases should be reported to governmental health agents or veterinary surgeons.

3. **Pseudorabies** (Aujeszky's disease) (mad itch) is an acute, frequently fatal herpes virus disease affecting pigs (the primary host), canids, cattle, rodents and many other species of animals. It is characterised by intense itching, mutilation and death in severe cases. Many pigs survive and remain carriers. Rats and mice also serve as carriers and reservoirs of the virus. Sudden death of pigs under 3 weeks of age is common, with no clinical signs. Usually sneezing and coughing and a fever of 40.5°C precedes death. Virus spreads via nasal discharge and saliva into feed and watering devices. A vaccine is available. Strict sanitation and quarantine measures are needed for its control.

4. **Swine pox** is a viral disease spread by lice. On the skin small red areas occur which develop into papules. After a few days these form small pustules or vesicles which dry and form scabs surrounded by a raised inflamed zone. Mortality is low and the disease can be controlled by ridding pigs of lice.

Contagious Bacterial Diseases of Pigs

Bacterial diarrhoeas   Enteric diarrhoeal infections can be caused by a variety of organisms with varying results. Their control lies generally in careful sanitation, isolation of infected animals and treatment with broad spectrum antibiotics given either by injection or inclusion in feed or water.

1. **Enteric salmonellosis** can be either acute or chronic caused by *S. cholerae-suis* and *S. typhimurium*. Fever, debility and depression are usual symptoms. The chronic form in pigs displays persistent diarrhoea, loss of appetite and weight and development of a grossly distended abdomen. Control involves good sanitation and management and administration of antibiotics in feed and/or water.

2. **Necrohaemorrhagic enteritis** caused by *Clostridium perfringens, Type C* occurs in pigs during the first few days of life. Pigs develop diarrhoea and reddening of the anus and many die within 12 hours. Treatment is usually ineffective. Control depends on developing immunity in the dam which is transmitted passively to the young via colostrum.

3. **Colibaccillosis enteritis** due to *Escherichia coli* appears in baby pigs from young dams that have not developed immunity themselves. Consuming faeces from older sows stimulates the development of immunity in gilts. There are several other contagious *E. coli* diarrhoeal infections, usually occurring in very young baby pigs.

4. **Swine dysentery** (vibrionic dysentery, bloody scours, black scours) most frequently occurs in pigs 8 to 14 weeks old, although all ages can be affected. Usually the pigs pass loose stools that contain blood and mucous. In young weaned pigs, up to 90% to 100% might be affected with a 20% to 30% death loss unless antibiotics and electrolytes in water are administered. Death usually results from dehydration and a loss of electrolytes. Swine dysentery is caused by an anaerobic bacterium, *Treponema hyodysenteriae*. Control depends on strict sanitation, vaccination and antibiotics, reduction of stress and other good management practices.

Other contagious bacterial infections

1. **Swine streptococcal infections** (contagious). *Streptococcus suis* and *S. equisimilis* are 2 organisms that cause epidemic outbreaks of meningitis, septicaemia and arthritis in pigs of all ages, with fever, lameness, paralysis, some convulsions and endocarditis. Daily injection of penicillin and broad spectrum antibiotics are used in treatment of affected pigs.
2. **Leptospirosis** results primarily from the corkscrew shaped microorganism *Leptospira pomona* and two other strains (*L. icterohaemorrhagiae* and *L. canicola*). When transmitted to man it is referred to as the 'swineherd's disease'. Diseased animals may show a wide variety of symptoms including fever, anorexia, slowed growth, icterus, bloody urine, late pregnancy abortion and death. Leptospirosis is spread through infected ponds and feed contaminated by urine from infected animals, including cattle, dogs and rats. Control depends on vaccination, treatment with broad spectrum antibiotics, sanitation and isolation of carriers.
3. **Swine erysipelas** (diamond skin disease) is caused by *Erysipelothrix rhusiopathiae*. It is an acute septicaemic disease contagious also to man, turkeys, and perhaps some other animals, causing high fever, death, debilitating arthritis, diamond-shaped skin lesions, endocarditis and chronic ill health. Control and treatment depend upon a twice per year vaccination programme, the elimination of carriers and antibiotic treatment of all sick animals; penicillin is the drug of choice.
4. **Swine plague** results from *Pasteurella susceptica* and *P. multocida* that cause a contagious respiratory infection, producing haemorrhagic septicaemia and pneumonias with high mortality in some outbreaks. Control is possible through vaccination programmes with multocida bacterins, sanitation, isolation and antibiotic treatment such as tetracycline in feed or water.
   Other contagious respiratory diseases of pigs are **mycoplasma pneumonia** and **pleuropneumonia** (*Haemophilus parahaemolyticus*) which are controlled in the same manner as swine plague.
5. **Atrophic rhinitis** due to *Bordetella bronchiseptica* inflames the mucous membranes and destroys the scroll-like turbinate bones in the nose. Other bacteria such as *Pasteurella* may intensify the disease. Viruses

may produce other forms of rhinitis. *B. bronchiseptica* can be carried in the respiratory tracts of many other mammals. Sneezing is the most common symptom accompanied by snuffing, snorting, coughing and excessive lacrimation. Twisting and/or shortening of the nose may occur. Pneumonia is the usual cause of death. The most severe lesions occur in pigs infected during the first few weeks of life. Vaccination, tetracyclines in feed and water and sanitation are means of control. It might be eradicated by slaughtering all animals, cleaning the premises and starting over with clean animals.

6. **Tetanus** is an infection from the soil-borne organism *Clostridium tetani*, that can be introduced into the body through puncture wounds, castration and other wounds. Individuals can be treated with tetanus antitoxin and penicillin injections. In herds on premises with a history of infection, vaccination with a tetanus toxoid is advisable.

Parasitic Diseases of Pigs
External parasites

1. **Lice pediculosis** (lousiness) is due to *Haematopinus suis*, a blood sucking parasite. It can be controlled with dips or with sprays of coumaphos (0.06%), lindane (0.03%), or with sprays of malathion (0.5%). Ivermectin is also effective.
2. **Mange**
   a. Sarcoptic mange from *Sarcoptes scabiei* var. *suis* causes intense itching, spreading from the head over the body to the tail and legs. Effective control is possible by spraying or by dipping with lindane at concentrations of 0.05% to 0.1%, or malathion (0.05%). Ivermectin (300 micrograms per kg body weight) administered by injection or mouth is also effective.
   b. Demodectic mange also occurs in pigs infected with mites that have become established deep in the hair follicles. The treatments for sarcoptic mange are not effective, but demodectic mange can be controlled by subcutaneous injection of ivermectin at 300 micrograms per kg body weight.

Internal parasites

1. **Coccidiosis** can be caused by 8 species of *Eimeria* (protozoa) and 1 each of *Isospora* and *Cryptosporidium*. Symptoms include haemorrhagic diarrhoeas, weight loss, and in severe cases, death. Control comes through sanitation and the use of sulphas or amprolium in the feed.
2. **Stomach worms** (3 types: *Hyostrongylus, Ascarops and Physocephalus*) cause diarrhoeas, weight loss and debilitation. Effective anthelmintics are thiabendazole, levamisole, dichlorovos and ivermectin.
3. **Ascaris** (large roundworms). The larvae *Ascaris suum* migrate through liver and lungs causing haemorrhage, fibrosis, pulmonary oedema and pneumonia. They reach 30 cm or more in length as adults in the intestine. Treatment includes piperazine preparations that are effective and

inexpensive. Levamisole is an effective, broad spectrum anthelmintic that can be given in feed or drinking water. Ivermectin is also effective when given orally or by injection.

4. **Thornyheaded worms.** *Macracanthorhynchus hirudinaceus* are giant intestinal worms (30 cm long) in the small intestine. They are transmitted by grubs or beetles in the soil such as in permanent pig pastures. Ploughing and cleaning premises using concrete feeding slabs, etc., help in control. Levamisole is an effective treatment.

5. **Giant kidney worms** (20 to 40 mm long). *Stephanurus dentatus* spread from eggs and larvae in urine via the soil and feed and water being eaten by pigs. The larvae can also penetrate the skin and migrate through the body to the kidneys. Ivermectin (300 micrograms per kg body weight) is the treatment of choice.

6. **Intestinal threadworms** (*Strongyloides*), **intestinal worms** (*Oesophagostomum*), **pinworms** (rectal worms of pigs, *Trichuris*), **lungworms** (*Metastrongylus elongatus*), and **lung flukes** (*Paragonimus westermanii*) are best treated with ivermectin 300 micrograms per kg body weight given subcutaneously by injection.

An effective worm control programme for pigs would include good hygiene and anthelmintics in the feed as follows:

1. Treat sows and gilts 5 to 7 days before farrowing and again in mid-lactation.
2. Treat weaner pigs prior to entering fattening pens, and again 8 weeks later.
3. Treat fattening pigs at 50 kg body weight.
4. Treat all boars at 6-month intervals. Alternate this in-feed treatment with ivermectin injections which are effective against both internal and external parasites.

MANAGEMENT

As in any operation, pig management in the tropics can be improved to obtain efficient production. However, this improvement in management need not entail only copying practices used in the temperate zone, but can include techniques peculiar to the special situation.

Environment

The pig is a non-sweating animal largely adapted to a warm, shaded, humid environment. The newborn pig is incapable of protecting itself from either excessive cold or heat. Thus, during the first 2 days of life the temperature around the baby pigs should be slightly over 32.2°C, the temperature being gradually lowered thereafter. Chilling produces lethargy and piglets have a more difficult time getting out of the way when the sow lies down, thus becoming more susceptible to being crushed by the sow. Ideal temperature for pigs weighing 32 to 65 kg is 24°C, for those weighing 75 to 118 kg 15.6°C. At temperatures of 32.2°C and above, respiration rates increase to 150 to 200 per minute, pigs stop eating and lose weight, and if forced to exercise may die of heat exhaustion. This would suggest that in the tropics with a mean annual temperature of about 26.7°C pigs should be raised only to 54 to 64 kg instead of the heavier 91 to 109 kg weight that is practised in temperate climates.

To alleviate heat stress pigs should be provided with shade and fine water sprays or wallows that are easily cleaned.

One of the most suitable and yet cheapest pens is one that is half covered by a roof providing shelter if desired. The roof should slope from a height of 1.4 to 3 m down to 1.8 to 2.1 m. It can be made of thatch (coconut fond, nipa, reed, grass, etc.) or of galvanised iron that could be painted aluminium on top and black beneath to further improve the microclimate underneath. Mature sows should have 1.67 to 1.86 sq m shelter or sleeping space each.

Boars and gilts can be reared together until about 4 months of age, at which time they should be separated. Then, only animals of approximately the same age should be held together. Boars tend to become vicious if penned alone for extensive periods; rather they should run with other boars or with pregnant sows.

Farrowing

1. About 36 hours before farrowing the vulva swells and the teats harden. Once parturition begins, piglets are normally born every 25 minutes. Assistance may be necessary if longer than 30 minutes elapses between birthings. Parturition lasts from 1 to 12 hours; the first piglet may suckle before the last piglet is born.
2. At farrowing the newborn pigs need protection from being crushed by the sow; crushing is responsible for the greatest loss of newborn pigs. Therefore, one week before farrowing, the pregnant sow should be confined in a farrowing crate so she can adjust to the circumstances. The farrowing crate should have guard rails extending about 15 cm out from the wall and be up to 20 cm from the floor. The crate should be long enough to comfortably hold the sow and narrow enough to prevent her from turning around. Overall dimensions should be 2.1 m long and 56 cm wide. Another crate design would be provided with guard rails, but would be large enough to allow the sow to turn around; this pen should be slightly longer than wide so that the sow would dung in a corner away from the feed trough.
3. A separate comfort zone for the baby pigs provides additional protection by attracting them away from the sow except when nursing.
4. Piglets start suckling immediately after birth, and after a few hours of jostling, each piglet has found his own teat that he keeps. The more vigorous piglets have claimed the most productive teats for their own. There is no point in trying to keep more piglets with a sow than there are functional teats. During the first 2 or 3 days of life, if they have received colostrum, surplus piglets can be transferred to other sows in farrowing crates. (If free moving, sows will not foster piglets not her own.) Newborn pigs weighing less than the average have a reduced

likelihood of surviving; about 60% of the newborn pigs weighing less than 900 g can be expected to die unless given special enriched milk and attention.

Solid creep feed should be introduced when the piglets are 1 week of age and continue until weaning at 6 to 12 weeks. Piglets are born with little body reserve of iron, and milk does not provide sufficient iron to meet their requirements; therefore, to prevent anaemia 3 days after birth, the piglet should be injected in the ham muscle with 100 mg of iron in the form of iron dextran. Lacking this, the sow's udder can be daily painted with a solution of 500 mg ferrosulphate and some sugar dissolved in 4 litres of water; or, fresh clean soil, free of parasite eggs and pathogens, can be placed where it is available to the sow and to her baby pigs.

5. Pigs, when born, have 8 small sharp tusk-like teeth (needle teeth) that should be clipped to prevent injuring the sow's udder or themselves. The tips of these teeth should be clipped at 3 days of age with a pair of side-cutting, toenail clipper pliers or special forceps. Remove about one-half of each tooth.

6. Baby pigs should be identified by ear notching before 6 weeks of age. The universal ear notching system involves both ears; the right is used to indicate the litter number (all the pigs in the same litter will have the same notches) and the left ear the pig number within the litter (each pig in the litter will have a different number) (Figure 55).

Each ear is divided into 4 areas (quadrants). The proximal (closest to the head) and distal (furthest) parts are again divided into a top and bottom, and a number is ascribed to each quadrant: one to the lower proximal, 3 to the lower distal, 9 to the upper distal and 27 is given to the upper proximal sections. Two notches in the same quadrant doubles its value, for example two notches in the lower ear closest to the head would represent 2. The tip is reserved for 81. The pig and litter numbers are determined by adding the values in the left and right ears, respectively.

Excess males should be castrated 7 to 20 days before weaning. If piglets are healthy, iron shots, ear notching, teeth cutting and castration can all be done at one handling at 3 to 6 days of age. At weaning, to reduce stress on the young pigs, the sow should be removed from the pigs, not the pigs from the sow; the pigs thus remain longer in familiar surroundings. Reducing the sow's feed and water at weaning will decrease the milk produced and udder congestion. Three to 7 days after weaning the sow will return to heat, and if in thrifty condition, should be bred.

7. Pigs at birth weigh about 1 kg, but they gain rapidly. Under optimal conditions the weight doubles in the first week and triples by the end of the second week. After 3 weeks, the weight might be 6.8 kg, and a market weight of 100 kg might be reached by 5 months of age. These values are attained by feeding about 3.56 kg of feed for every kg of gain.

8. The most efficient gains are made when the pig is young and growing.

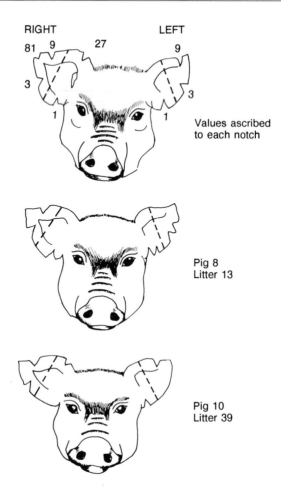

Figure 55 *Ear notching system for identifying pigs. Notches in the right ear identify the litter and those in the left ear designate the individual pig within the litter.*

This is because of the nature of the body weight increase. The whole carcass composition might be:

Body weight	H_2O	Protein	Fat	Ash
8 kg	73	17	6	3.4
30 kg	60	13	24	2.5
100 kg	49	12	36	2.6

As with other animals, the composition of the body changes as the body weight increases (concurrent with increased age). Changes occur in the composition of both the total carcass and in the body tissue that is actually added. As seen in Figure 15, p. 43, as the animal becomes heavier (older), the total protein and ash in the carcass slightly increases. The total water in the body increases rapidly until the animal approaches maturity at which time the fat increases at the most rapid rate. As the body increases in weight, the proportion of this added weight as ash, protein, fat and water can be shown as in Figure 15B. As the animal becomes heavier, the protein proportion of any gain declines only slightly. However, the fat increases drastically at the expense of water.

The overall efficiency with which feed energy is deposited in growth is about 74%; in young pigs the efficiency with which tissue is deposited is about 80% for protein and 64% for fat. In older animals these values become 52% and 70% for protein and fat, respectively. This suggests that the efficiency of depositing protein declines whereas that of depositing fat increases with increased size (age). Relationships depicted in Figure 15 emphasise that younger, faster growing animals make the most efficient use of their feed energy.

Pigs as Scavengers

Pigs can supplement their intake by scavenging. The difficulty encountered in raising pigs on the ground and out of doors is the high incidence of internal parasites. The producer should make periodic use of anthelmintics that would be effective against the local infestations.

To improve the productivity of scavenging pigs the small scale farmer could:

1. Supply supplementary feeds such as rice bran and peelings of root crops (preferably cooked) once or twice a day in an area adjacent to the house. A difficulty might be in keeping the neighbouring pigs away from the feed.
2. If land and sufficient feed are available, paddocks adjacent to the house could be fenced with either wire netting, plaited bamboo, paling wood (pickets) or a closely planted live-fence species. The paddock should be subdivided into 4 to 6 smaller areas so that the pigs could be moved from one enclosure to another at 10 to 14 day intervals. This would both help preserve the ground cover and reduce the incidence of parasitic infection. While this reduces the scavenging possibilities, it does increase the control of animals and permits the mating of females to selected boars and thus improving the stock.
3. If the resources (including feed) allowed, pens could be built with a cover and a sloping concrete floor that could be cleaned. Pigs could be trained to deposit their faeces at the lower end of the pen. Sires of improved breeds could be used to upgrade as quickly as and to the extent that the resources and management could justify.

Estimating Pig Weights

The weight of a pig can be estimated by measuring the heart girth and body length from poll to base of tail and applying the formula:

$$\frac{(\text{heart girth, inches})^2 \times (\text{body length, inches})}{400} = \text{body weight in pounds}$$

In metric terms the weight in kg can be approximated by applying the formula:

$$\frac{(\text{heart girth, cm})^2 \times (\text{body length, cm})}{14,440} = \text{body weight in kg}$$

Record Keeping

Successful management would entail keeping detailed records of such items as breedings, farrowings, individual pig identification through ear notching, growth response, disease outbreaks, rations fed, etc.

Farmers entering a hog producing project should make a list of personnel and other resources available including sources of feed, breeding animals, veterinary supplies and services that can be drawn upon as problems might develop.

APPENDICES

GLOSSARY

Abomasum   In the ruminant stomach, the fourth and true digestive compartment. The forestomach consists of the rumen, reticulum and omasum.

Abortion   Premature expulsion from the uterus of an embryo or non-viable foetus.

Abscess   A localised collection of pus surrounded by inflamed disintegrating tissue.

Acidosis   An accumulation of acids in the body upsetting the acid-base balance.

Active transport   Movement of molecules across a membrane against a concentration gradient with the help of energy.

Acute   Sharp; having a short and relatively severe course.

Adhesion   Abnormal union of adjacent tissues.

Ad libitum (ad lib.)   Free choice, or at one's pleasure.

Aetiology   Study of the causes of disease.

Agalactia   Failure to secrete milk.

Alkaloid   One of many plant nitrogen-containing ring compounds, bitter in taste and having abnormal physiological effects on the body (e.g. nicotine, cocaine and morphine).

Alkalosis   A loss of acid (CO_2) in the body, possibly due to hyperventilating lungs. The blood becomes abnormally alkaline.

Allele   One of a multiple of forms of specific genes located at the same locus of homologous chromosomes.

Alveolus (pl., alveoli)   A tiny functional sac at the end of a tubule in a gland or organ (e.g. lung, udder). Any of a group of possible mutational forms of a gene.

Amino acid   An organic compound possessing a basic amino ($-NH_2$) group and an acidic carboxyl ($-COOH$) group; the building blocks for protein.

Anaemia   Abnormal decrease in haemoglobin and/or red cells in the blood resulting in oxygen transport failure.

Anaerobic   The absence of air. Anaerobic organisms grow in the absence of air or oxygen.

Anaphylaxis   An exaggerated reaction to a foreign protein; an antigen-antibody reaction.

Anatomy   The structure of an organ or organism. Physiology deals with the functioning of an organ or organism.

Anorexia Loss of appetite for food.

Anthelmintic (or antihelmintic) An agent that is destructive to worms.

Antibiotic Chemical substance produced by microorganisms which has the ability to inhibit the growth of, or to destroy, other organisms.

Antibody A type of protein (modified gamma-globulin) in the blood that has been produced as a consequence of an encounter of the immune system with an antigen. It opposes the action of disease organisms.

Antigen A compound (usually a protein molecule or organism foreign to the body) that induces the formation of antibodies.

Arthritis Inflammation of a joint.

Atom The smallest unity of an element that can exist and still retain the chemical properties of the element. It consists of a centre positively charged nucleus surrounded by negatively charged electrons. The number and arrangement of these electrons determine all the properties of the atom except its atomic weight and radioactivity.

Attentuate To render a pathogen or toxin less virulent and destructive.

Autonomic Self-controlling; acting without outside direction.

Bactericide An agent that destroys bacteria.

Bacterin A suspension of killed organisms used to immunise against a specific disease.

Ballottement (bumping) A feeling manoeuvre to test for a floating object such as a foetus in the abdominal cavity.

Biologic A term to include such medicinals as hormones, antibiotics, sulfas and other drugs administered to increase production.

Bolus A mass of food ready for swallowing (or for regurgitating in rumination).

Broiler A chicken raised for meat.

Broody Prolactin-induced maternal instinct in hens to cease egg laying and to hatch a clutch of eggs.

Browse Twigs, shoots and leaves of forbs, shrubs and trees.

Buck Mature male rabbit or goat.

Calcinosis A deposition of calcium salts in various tissues of the body.

Calorie A measure of energy, more specifically of heat. A calorie is defined as the amount of heat required to raise the temperature of water 1°C (specifically from 14.5°C to 15.5°C). The calorie is equivalent to 4.1855 joules.

Cannibalism Eating its own kind (species).

Castration Removal of the testes.

Cavernous Containing hollow spaces which fill with blood under pressure to produce erection as in the penis.

Chevon The meat of the goat.

Chromosome One of a pair of dark-staining threadlike bodies that appear in pairs at cell division carrying the genes. One of each pair goes to each sex cell (gamete) in spermatogenesis or oogenesis.

Chronic Persisting over a long period of time.

Climate The composite prevailing weather conditions of a region.

Cloaca The chamber that receives the discharges of the digestive, urinary and reproductive tracts of birds. It is closed by the vent.

Clutch   A nest of eggs laid in a certain time pattern.

Cockerel   Young male chicken; immature cock or rooster.

Cofactor   A non-protein substance that is an essential participant in an enzyme system. Some vitamins and metallic ions function as cofactors.

Collogen   The main supportive, fibrous protein of the skin, tendon, cartilage, bone and connective tissue. When boiled it is converted into gelatin.

Colon   The intestine, usually the large intestine or the part extending from the caecum to the rectum.

Colostrum   The secretion of the mammary gland just before and after parturition. It is rich in antibodies, protein, vitamins and minerals.

Compound   A pure substance consisting of 2 or more elements in definite proportions usually having properties unlike those of its constituent elements; glucose and lysine are examples .

Conception   Union of the sperm and ovum (fertilisation) in the formation of a viable zygote expected to develop into an embryo and a new individual.

Conceptus   The whole product of conception, the offspring, accompanying membranes, and fluids.

Contagious   Capable of being transmitted from one animal to another.

Coprophagy   Feeding on dung or excrement, or eating its own faeces.

Copulation   Sexual union of two individuals with the disposition of sperm in the vicinity of the ovum.

Creep   An enclosure allowing the young but not the mature animals to enter and eat extra feed.

Criollo   A native breed or strain of livestock derived from animals introduced from outside the area.

Criterion   A standard or basis of judgement.

Crypt   A minute, tubelike depression opening on a free surface.

Cryptorchidism (cryptorchism)   Failure of one or both testes to descend from the abdomen into the scrotum.

Cyanosis   A bluish discoloration of the skin and mucous membranes resulting from inadequate oxygenation of the blood or from a restricted blood supply to a part.

Dam   A female parent.

Dermatitis   Inflamation of the skin, a symptom of a number of nutrient deficiencies as well as of infections, allergies and chemical irritations.

Detergent   Any of a group of water-soluble cleaning agents that have wetting and emulsifying properties.

Diarrhoea   Abnormal frequency and liquidity of faecal discharges; dysentery or scours.

Digestion   The changes taking place in feed as it traverses the digestive tract in preparation for absorption and use by the animal body.

Disease   An abnormal condition of an organism or part that impairs normal functioning. It could be due to infection, inherited weakness or environmental stress.

Disinfectant   A chemical agent that destroys microorganisms but not bacterial spores.

Doeling   A diminutive term referring to a young doe.

Duodenum   The upper segment of the small intestine into which the contents of the stomach and secretions of the liver and pancreas empty.

Dystocia   Abnormal labour or birth due principally to unusual size, shape or position of the foetus in relation to the size and shape of the maternal pelvis.

Ecology   Branch of biology dealing with the relations between organisms and their environment.

Ejaculate   Sudden act of ejecting semen; discharge.

Electrolyte   A chemical compound that will conduct electricity when dissolved in a solvent such as water, or when molten. In body fluids electrolytes consist principally of sodium (extracellular) and potassium (cellular) chlorides and bicarbonates. Other chemicals in much smaller amounts are calcium, magnesium, sulphate and phosphate.

Element   A fundamental constituent of something; a simple substance that cannot be decomposed by chemical means. There are 103 elements that are recognised, carbon (C), hydrogen (H), and oxygen (O) being examples.

Embyro   Early (first trimester) stage of an animal in which all the different tissues of the adult are developed; it then becomes a foetus.

Emulsion   A dispersion of liquids that do not ordinarily mix, like lipid and water, as with milk fats in milk.

Endemic (enzootic)   Said of a disease that is constantly present in an area, but usually affecting only a small number of animals at any one time.

Endocrine   Applied to ductless organs secreting hormones into the bloodstream to be carried elsewhere in the body to affect a function.

Endotoxin   A heat-stable poison present within bacterial cells. It is not liberated until the cell dies and disintegrates. Endotoxins produce fever and increase capillary permeability.

Energy   Capacity to do work; power to produce motion and to effect changes. Different manifestations of energy include mechanical, chemical, heat, electrical and radiant. The calorie, though actually being a unit of heat, is used to measure energy in nutrition. The joule has wider applications in measuring energy.

Enteric   Pertaining to the intestine.

Enteritis   Inflammation of the intestine, especially the small intestine.

Enterotoxaemia   Presence in the blood of toxins (poisons) absorbed from the intestine. Toxins are produced by organisms such as *Clostridium perfringens, C. spiroforme* and *E. coli.*

Environment   The sum total of surrounding things, conditions or influences. This includes feed, disease, climate, care and keeping, protection, exposure, etc.

Enzyme   A protein produced by a living organism and functioning as a biochemical catalyst in living tissues. Enzymes are specific for particular chemical structures or types of substrates.

Epithelium   The covering of the external and internal surfaces of the body.

Equilibrium A state of balance; the acid-base equilibrium is a normal ratio between the acid and basic elements of the blood and body fluids.

Eructation Belching; the release of gas from the stomach.

Essential amino acids Amino acids essential to metabolism that cannot be formed in the body and thus must be included in the diet.

Exotic Not indigenous to an area; foreign, unusual.

Exudate Fluid, cells, bacteria and other debris escaping the tissues usually in consequence of an infection.

Faeces The waste excreted from the intestine consisting of bacteria, cells shed from the intestinal lining, secretions from the liver and mucus, and feed residues. The lower the digestibility of the feed, the greater the feed residues.

Farrier Pertaining to horseshoeing; one who shoes horses; a blacksmith.

Farrowing Sow giving birth to her young.

Fatty acid A molecule with a carbon and hydrogen chain and a carboxyl (COOH) group on one end; fats are made up of 3 fatty acids and glycerol.

Fecundity Prolificity; ability to produce offspring rapidly and in large numbers.

Fibrosity Pertaining to both the fibre content and coarseness of a feed; relative to the inherent requirement of some animals for large particles in the diet.

Flushing Providing extra feed for an animal so that it will gain weight, usually prior to breeding.

Fodder The entire above-ground mature feed plant including the stalk, leaves and ears. The term may apply to any dry roughage or forage.

Foetus The unborn in the last 2 trimesters of gestation; the embryo develops into a foetus.

Follicle A group of cells containing a cavity or small sac, as in the ovarian or hair follicles.

Forage Vegetative part of plants used for animal feed.

Forb A non-grass weed or broadleaf flowering plant whose stem above ground does not become persistent or woody.

Founder Lameness caused by inflammation of the laminae between the hoof and coffin bone, usually resulting from overeating high energy feeds.

Freshen To give birth to young. This initiates lactation in mammals.

Fryer A young animal such as a chicken or rabbit customarily prepared by frying.

Gastric Pertaining to the stomach.

Gene The biologic unit of heredity situated on a chromosome. It is composed of deoxyribonucleic acid (DNA).

Genetics The study of heredity.

Genotype The genetic makeup of an animal; its heredity.

Gestation Period of development of the young in the uterus from fertilisation to birth.

Gilt A young female pig that has not yet produced a litter.

Gonad A sex gland in which the sperm or ova are produced; testes in the male, ovary in the female.

Gossypol An aromatic aldehyde produced in the pigment glands of cottonseed. Its toxic effect can be reduced by feeding extra iron.

Haemorrhage A copious escape of blood from a vessel; bleeding.

Helminth A worm or wormlike parasite.

Hepatitis Inflammation or disease of the liver.

Herbivore An animal subsisting on plant matter for feed.

High energy bond A chemical bond representing an unusually high amount of energy used in metabolism. Bonds holding the phosphate radicals on ATP are the best known.

Homeostasis The tendency to uniformity or stability. A state of physiological equilibrium produced by a balance of functions and of chemical composition within the body. The desired physiological state.

Homologous Corresponding as pairs of similar chromosomes, one from the sire and one from the dam.

Hormone A substance produced by an endocrine gland to exert a specific effect on a distant target part of the body.

Hutch A cage for rabbits.

Hybrid Offspring of 2 animals of different breeds or species.

Hybrid vigour The increase in fitness and/or productivity of offspring over their parents due to heterozygosity. The increased vigour is most pronounced in the first generation crossbreeding and results largely from increased feed consumption.

Hydrochloric acid (HCl) The strong acid secreted by the glands of the stomach to assist in digestion. Also called muriatic acid.

Hydrolysis Chemically splitting a compound by taking up the elements of water. It is the principal chemical action of digestion.

Hygiene The science of health and of disease prevention.

Hypoglycaemia An abnormally low glucose content in the blood.

Hypothalamus A body part located at the base of the brain that is intimately associated with the pituitary; working together they regulate many endocrine glands and autonomic nerves controlling a variety of factors including temperature, blood pressure, rage, pain, regulation, visceral functions (e.g. hunger and thirst) and reproductive behaviour.

Icterus Jaundice; the yellowing of the mucous membranes and skin due to an increased level of bile pigments in the blood.

Immunity Resistance; the capability to resist and overcome infection or disease.

Impaction Condition of being firmly lodged or retained.

Incisors Front teeth used in cutting and grasping.

Increment The amount by which a given quantity is increased or decreased.

Indigenous Native, or living naturally in an area.

Infection Presence of pathogenic microorganisms within the tissues.

Infestation Invasion of the body by arthropods, including insects, mites and ticks; being overrun by vermin.

Ingesta Food and drink that has entered the body through the mouth.

Ingestion The act of taking something into the body; eating.

Inter Between, among, in the midst of, reciprocally.

Interaction   Different factors combining to produce an effect; the phenomenon wherein the combined effect of two factors is different from the simple sum of the responses to each factor.

Intra   Within, in, inside of.

Ion   An atom or group of atoms having an electrical charge. Cations carry positive and anions carry negative charges.

Joule   A measure of energy. The work done when the point of application of a force of 1 newton is displaced a distance of 1 metre in the direction of the force. The joule can measure mechanical and electrical energy as well as heat. One joule is equivalent to 0.2389 calories.

Ketosis   Accumulation of ketones in the body due to incomplete metabolism of fats, usually associated with a deficiency or inadequate utilisation of carbohydrates. Ketones are acid in nature.

Kidding   Process of parturition in the goat.

Kindling   Process of parturition in the rabbit.

Kit   A young rabbit from birth to weaning.

Lacrimation   Production of tears.

Lactation   Secretion of milk.

Lactic acid   2-hydroxypropanoic acid produced in the metabolism of glucose ($CH_3CHOHCOOH$).

Libido   Sex drive or desire.

Litter   The number of young brought forth at one birth in the cat, dog, sow, rabbit or other multiparous animals.

Locus (pl. loci)   Specific place (site) of a gene in a chromosome.

Lumen   The cavity or channel within a tubular gland or organ.

Mammal   Any vertebrate that feeds its young with milk from the female mammary glands.

Mastitis   An inflammation of the mammary gland. Mammitis.

Medial   Extending toward the middle or midline.

Membrane   A thin layer of tissue covering or separating structures or organs.

Metabolism   Total of all processes of life. It is the sum of all the physical and chemical processes by which living organised substance is produced and maintained, and also the transformation by which energy is made available for the uses of the organism .

Metritis   Inflammation of the uterus or womb.

Micoplasma   A type of infective organism.

Microfauna   Microscopic animal life which is characteristic of a special location.

Microflora   Microscopic plant life present in a special location.

Mol   The molecular weight of a substance in grams.

Molecule   A combination of 2 or more atoms which form a specific chemical substance.

Morbidity   Pertaining to sickness or disease; the sick rate or ratio of sick to well animals.

Mortality   Quality of being subject to death; the death rate; the ratio of number of deaths from a disease to the number of cases of the disease.

Moult   A periodic shedding of outer covering (i.e. feathers or fur) to be replaced by new growth.

Mycotoxin A poisonous substance produced by mould in feed that causes sickness or death.

Necropsy A post mortem (after death) examination of an animal carcass to learn the cause of death.

Nematode Any of the unsegmented worms having cylindrical, elongated bodies. Roundworms of the Ascaridae and Strongylidae families are among the most important internal parasites.

Nitrogen-free-extract (NFE) In the proximate analyses NFE represents the difference obtained when the percentage of moisture, protein, fat, mineral and fibre are all subtracted from the sample total (100%). NFE is thought to be the more digestible carbohydrate in the feedstuff although it includes the hemicellulose and part of the lignin.

Nucleus (1) The body within a living cell that contains the cell's hereditary DNA and produces the RNA that controls its metabolism, growth and reproduction. (2) The positively charged central region of an atom.

Nutrient Chemical compounds and elements contained in feeds used to support maintenance, growth and vital processes and to provide energy.

Oedema Accumulation of fluid in the intercellular spaces of body tissues causing swelling.

Oestrus A period and/or behaviour of the female mammal in which she is receptive to mating.

Omnivore An animal capable of subsisting upon food derived from both plants and animals.

Oocyst A reproductive cell inside a cyst.

Open A term applied to non-pregnant females.

Oral Pertaining to the mouth.

Osmosis The passage of a pure solvent from a lesser to a greater concentration when 2 solutions are separated by a semi-permeable membrane (one that allows passage of the solvent but not the solute).

Osmotic pressure The force that a dissolved substance exerts on a semi-permeable membrane. A high osmotic pressure will draw liquids from surrounding tissue.

Oxidation A chemical reaction in which oxygen is added, hydrogen is lost or an electron is removed. The changes brought about by the addition of oxygen. The opposite of reduction.

Palatable Pleasing or acceptable to the taste.

Palpation Feeling with the hands. Manual examination, handling.

Papule A small, circumscribed, solid elevation of the skin.

Parasite An animal or plant living upon or within another at whose expense it obtains some advantage without giving compensation.

Parenteral Administering substances to an animal by means other than the alimentary canal.

Parturition Act of giving birth.

Pathogen Any microorganism or material that will produce a disease.

Pedigree Record of ancestry.

Peptide Compound yielding 2 or more amino acids on hydrolysis. Formed by the peptide linkage in which the amino (NH_2) group on one amino acid and the carboxyl (COOH) group on another are joined with the loss of water (H_2O).

Peristalsis Wavelike muscular contractions to move material along a tubular organ such as the intestine, oviduct, etc.

Peritoneum Transparent membrane lining the abdominal cavity.

pH Symbol used to express acidity and alkalinity. Generally, pH of 7 is neutral, below 7 acidity increases, above 7 alkalinity increases. pH is the negative logarithm of the hydrogen ion concentration.

Phenomenon An observable fact, occurrence or circumstance.

Phenotype What an animal looks like; the outward expression of the genotype.

Pheromone A hormonal material secreted by one individual that stimulates a physiological or behavioural response in another individual of the same species.

Photoperiod The interval in a 24-hour period during which an animal is exposed to light.

Physiological Pertaining to the functions of living animals and their parts. The term anatomical pertains to the physical form of an animal or a part.

Pituitary A hormone-secreting gland at the base of the brain. The hypophysis cerebri gland is divided into the adeno-hypophysis, the anterior portion, and is loosely referred to as the master gland, and the neuro-hypophysis, the posterior pituitary.

Placenta Afterbirth; an organ surrounding the foetus by which it is united with the uterine wall. It is a foetal sac derived from both the foetus and uterus.

Plasma In blood, the liquid portion in which the cells are suspended. Plasma comprises both the serum and fibrinogen.

Pneumonia Inflammation of the lungs.

Poll The part of the head between the horns or between the ears. The top part of the head that is normally occupied by the horns.

Polled Hornless.

Postpartum After or following birth.

Precipitate To cause a substance in solution to settle out.

Precursor A forerunner, something that goes before.

Predator An animal habitually preying upon others for food.

Pregnant Condition of having an offspring developing within the body.

Proenzyme An inactive form of an enzyme requiring the action of some other material to become active; a zymogen.

Progeny Offspring; descendants.

Protozoa Microscopic animals consisting of a single cell. Some perform useful purposes in the rumen whereas others, such as coccidia, are pathogenic.

Pseudopregnancy False pregnancy; the female may go through all the behaviour patterns associated with pregnancy during the entire gestation period.

Puberty Age at which an animal is first capable of reproduction.

Pullet A young female chicken.

Pulse (1) Regular throbbing of the arteries caused by successive contractions of the heart. (2) Seeds of leguminous plants; they are rich in protein but generally possess anti-nutritional factors such as protein inhibitors, gastrogens, cyanogens, antivitamins and lathyrogens. These harmful factors can often be inactivated by cooking, germination and/or fermentation.

Pus A thick yellowish-white fluid formed in infected tissue, consisting largely of leucocytes, cellular debris and liquified tissue elements.

Qualitative Relating to quality, composition, or makeup.

Quantitative Pertaining to quantities or amounts.

Ratchet A mechanism equipped with teeth into which a pawl engages, allowing movement in one direction only.

Reduction The chemical reaction in which oxygen is removed or hydrogen or an electron is added. The opposite of oxidation.

Reflex An automatic (involuntary) coordinated response to a stimulus.

Regression In animal breeding, the tendency for offspring to be like the average of the population rather than the average of the parents.

Respite A temporary cessation of a situation or condition that is uncomfortable or stressful.

Roaster A young meat animal more mature than a fryer.

Ruminant Even-toed, hooved mammal with compartmentalised stomach capable of regurgitating a bolus of feed (cud) for re-chewing.

Salivation The production of saliva, a watery liquid produced in the mouth to aid chewing and swallowing and, in omnivores, digestion of starch.

Saturated fat Glycerol ester of fatty acids that have no double or triple bonds; hard fat. Animal fats are generally more saturated than those from plants.

Scavenger An animal that searches for feed including dead organic matter.

Scorbitic Pertaining to scurvy or deficiency of vitamin C.

Scours Severe diarrhoea.

Septicaemia Invasion of the bloodstream with disease-producing organisms (pathogens).

Serum In blood, the clear liquid remaining after the more solid elements (cells and fibrin) have been separated.

Shrub A low, woody plant with several stems.

Solute A substance dissolved in a liquid to form a solution.

Solvent A liquid in which a substance (solute) is melted away (dissolved).

Sophisticated Changed from the natural character or simplicity to a more complex nature; not naive.

Sphincter A ring-like muscle surrounding a body opening, such as the anal sphincter.

Spore A usually single-celled reproductive structure or resting stage of a fungus, bacterium or protozoan.

Stanchion Upright attached device for closing on both sides of the neck to restrict the movement of an animal.

Starter A readily digested grain concentrate fed to kids prior to weaning.

Stover Roughage matter other than grain. The vegetative part of a plant (e.g. maize or sorghum) which remains after the ears or heads have been removed.

Straw The vegetative plant residue after seeds have been removed. Although usually used with regard to cereals the term might also apply to dry stalks and chaff of legumes.

Stress Any stimulus (fear, pain, discomfort) that disturbs the normal physiological equilibrium of an animal.

Striated Muscle fibres having parallel cross stripes, as in voluntary (skeletal) muscles.

Subcutaneous Under the skin.

Substrate The material acted upon by an enzyme.

Succulent Juicy; having a high moisture content.

Succumb Yield to disease, old age, etc.

Suppuration Formation and discharge of pus.

Symptom Any functional evidence of disease or abnormality.

Synthesis The process of forming with the input of energy a more complex compound from elements or simpler compounds.

Tetany Rigidity and spasmodic contraction of the voluntary muscles.

Tether A chain or rope by which an animal is tied to a fixed object (post, stake or tree).

Therapy Treatment of a disease or deficiency.

Thermoneutral That environmental temperature at which an animal does not need to work to maintain its body temperature.

Toxin A poison frequently referring to a protein with high molecular weight produced by a plant, animal or microbe capable of producing antigens and damage to body tissues.

Toxoid A toxin that has been modified so as to cause bodily damage no longer but that can still produce antigens that will create immunity; tetanus toxoid.

Transudation Passage of a body fluid through a membrane as in the movement of moisture to the surface of the skin.

Trauma Wound or injury.

Tropics Regions lying between and near the latitudes 23.5° North and 23.5° South.

Undulate To move with a wave-like motion.

Vaccinate To inject a suspension of attenuated or killed microorganisms or other biological materials for the purpose of producing immunity to an infectious disease.

Vasoconstriction A closing off of the capillaries restricting the blood flow to the tissues.

Vasodilation A state of increased calibre of the blood vessels, increasing the flow of blood to a body part.

Vent The term applied to the anus of a chicken. This single body opening serves the digestive, urinary and reproductive systems. The cloaca collects the products until voided through the vent.

Ventral Denotes a position more toward the belly surface or to a part of the body farthest from the vertebral column; opposite to dorsal.

Vermicide (or vermifuge) An anthelmintic drug destructive to intestinal animal parasites (worms).

Vermin Noxious or objectionable insects or small animals, some of which may be parasitic.

VFA Volatile, short-chain fatty acids, principally acetic, propionic, butyric and valeric.

Villi The microscopic projections that extend into the lumen; in the intestine they increase its surface for better absorption of nutrients.

Virulent Highly infective, malignant or deadly.

Viscera The organs in the thoracic, abdominal and pelvic cavities of the body, especially the abdominal.

Wry Distorted or lopsided.

REFERENCES

Anatomy and Physiology of Farm Animals, R. D. Frandson. Lea & Febiger, Philadelphia, PA. 1965.

Animal Agriculture, H. H. Cole and Magnar Ronning. W. H. Freeman and Company, San Francisco, CA. 1974.

Animal Husbandry in the Tropics, G. Williamson and W. J. A. Payne. Longman Inc., New York, NY. 3 ed. 1978.

Biology and Medicine of Rabbits and Rodents, John E. Harkness and Joseph E. Wagner. Lea & Febiger, Philadelphia, PA. 2 ed. 1983.

Biology of Earthworms, C. A. Edwards and J. R. Lofty. Bookworm Publishing Co., Ontario, CA. 1976.

Commercial Chicken Production, Mack O. North. AVI Publishing Co., Westport, CN. 2 ed. 1978.

Compendium of Rabbit Production, Appropriate for Conditions in Developing Countries. Wolfgang Scholaut. GTZ GmbH, Eschborn, Germany. 1985.

Dairy Goats, Breeding/Feeding/Management, B. E. Colby, et al. Agricultural Cooperative Extension Service, University of Massachusetts, Amhurst, MA. 1972.

Diseases and Parasites of Livestock in the Tropics, H. T. B. Hall, Longman Group Ltd., London. 1977.

Earthworms for Ecology and Profit. R. E. Gaddie, Sr. and D. E. Douglas. Bookworm Publishing Co., Ontario, CA. 1977.

Extension Goat Handbook, G. F. W. Haenlein and D. L. Ace, ed. Extension Service, USDA,Washington, D.C. 1984.

Feeds and Nutrition, M.E. Ensminger and C. G. Olentine, Jr. Ensminger Publishing Co., Clovis, CA. 1978.

Handbook of Livestock Management Techniques, R. A. Battaglia and V. B. Mayrose. Burgess Publishing Co., 7108 Ohms Lane, Minneapolis, MN 55425. 1981.

Merck Veterinary Manual, O. H. Siegmund, ed. Merck & Co., Rahway, NJ. 5 ed. 1979.

Nutrient Requirements of Goats, NRC #15, National Academy Press, Washington, D.C. 1981.

Nutrient Requirements of Poultry, NRC, National Academy Press, Washington, D.C. 1977 and 1984.

Nutrient Requirements of Rabbits, NRC, National Academy Press, Washington, D.C. 1977.

Poultry Science, M. E. Ensminger. Interstate Publishers, Danville, IL. 2 ed. 1980.

Practical Poultry Raising. Peace Corps Information Collection and Exchange, Washington, D.C. 20526. 1981.

Rabbit Production, P. R. Cheeke, N. M. Patton, S. D. Lukefahr and J. I. McNitt. Interstate Publishers, Danville, IL 6 ed. 1987.

Raising Small Meat Animals, V. M. Giammattei. Interstate Publishers, Danville, IL.1976.

Small Poultry Flock, C. C. Sheppard, et al. Michigan State University Extension Bul. 773. 1978.

Small Scale Poultry Production, Tokushi Tanaka. Cooperative Extension Circular 480, University of Hawaii. 1972.

Stockman's Handbook, M. E. Ensminger. Interstate Publishing Co., Danville, IL. 6 ed. 1983.

METRIC CONVERSION TABLES

BRITISH TO METRIC   METRIC TO BRITISH

Length

1 inch (in)	= 2.54 cm	1 millimetre (mm)	= 0.0394 in
	or 25.4 mm	1 centimetre (cm)	= 0.394 in
1 foot (ft)	= 0.30 m	1 metre (m)	= 1.09 yd
1 yard (yd)	= 0.91 m	1 kilometre (km)	= 0.621 miles
1 mile	= 1.61 km		

Area

1 sq inch (in^2)	= 6.45 cm^2	1 sq centimetre (cm^2)	= 0.16 in^2
1 sq foot (ft^2)	= 0.093 m^2	1 sq metre (m^2)	= 1.20 yd^2
1 sq yard (yd^2)	= 0.836 m^2	1 sq metre (m^2)	= 10.8 ft^2
1 acre (ac)	= 4047 m^2	1 hectare (ha)	= 2.47 ac
	or 0.405 ha		

Volume (liquid)

1 fluid ounce (1 fl oz)		100 millilitres (ml or cc)	= 0.176 pints
(0.05 pint)	= 28.4 ml	1 litre	= 1.76 pints
1 pint	= 0.568 litres	1 kilolitre (1000 litres)	= 220 gal
1 gallon (gal)	= 4.55 litres		

Weight

1 ounce (oz)	= 28.3 g	1 gram (g)	= 0.053 oz
1 pound (lb)	= 454 g	100 grams	= 3.53 oz
	or 0.454 kg	1 kilogram (kg)	= 2.20 lb
1 hundredweight (cwt)	= 50.8 kg	1 tonne (t)	= 2204 lb
1 ton	= 1016 kg		or 0.984 ton
	or 1.016 t		

U.S. TO METRIC

1 teaspoon	= 5 ml
1 tablespoon	= 15 ml
1 fluid ounce	= 30 ml
1 cup	= 0.24 litre
1 pint	= 0.47 litre
1 quart	= 0.95 litre
1 gallon	= 3.8 litres

METRIC TO U.S.

1 litre	= 200 teaspoons
1 litre	= 66.7 tablespoons
1 millilitre	= 0.03 fluid ounce
1 litre	= 4.2 cups
1 litre	= 2.1 pints
1 litre	= 1.06 quarts
1 litre	= 0.26 gallons

Temperature

$(°C \times 1.8) + 32 = °F$

$(°F - 32) \div 1.8 = °C$

INDEX

INDEX

Italicised numbers refer to illustrations

FARMING PRESS BOOKS

Below is a sample of the wide range of agricultural and veterinary books published by Farming Press. For more information or for a free illustrated book list please contact:

Farming Press Books, 4 Friars Courtyard
30–32 Princes Street, Ipswich IP1 1RJ, United Kingdom
Telephone (0473) 241122

Goat Farming ● ALAN MOWLEM
Covers all aspects for those considering goats as an alternative enterprise.

All About Goats ● LOIS HETHERINGTON
An introduction covering feeding, milking, kidding, housing, rearing, management, health, products and showing.

Goatkeeper's Veterinary Book ● PETER DUNN
Diagnosis, treatment and prevention of goat ailments.

Free-Range Poultry ● KATIE THEAR
Non-intensive poultry keeping for both small-scale and larger-scale units.

Outdoor Pig Production ● KEITH THORNTON
How to plan, set up and run a unit.

Calculations for Agriculture & Horticulture ● GRAHAM BOATFIELD &
IAN HAMILTON
Provides the calculation methods for crops, livestock, machinery and horticulture.

Pigs in the Playground ● JOHN TERRY
A humorous account of the transformation of wasteland to a successful school farm.

Farming Press Books is part of the Morgan-Grampian Farming Press Group which publishes a range of farming magazines: *Arable Farming, Dairy Farmer, Farming News, Pig Farming, What's New in Farming*. For a specimen copy of any of these please contact the address above.